The Comparative Endocrinology
of Calcium Regulation

The Comparative Endocrinology of Calcium Regulation

Edited by
C Dacke, J Danks, I Caple and G Flik

A publication of the Society for Endocrinology

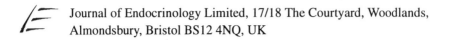

Journal of Endocrinology Limited, 17/18 The Courtyard, Woodlands, Almondsbury, Bristol BS12 4NQ, UK

A publication of the Society for Endocrinology
© 1996 Journal of Endocrinology Limited

British Library Cataloguing in Publication Data
A CIP catalogue record for this book is available from the British Library.

ISBN 1 898099 08 1

Cover design by Eat Cake Design
Printed in Great Britain by Bourne Press

The XIIth International Conference
on
Calcium Regulating Hormones

14-19 February 1995, Melbourne, Australia

Preface

This book contains collected papers from the meeting on the Comparative Endocrinology of Calcium Metabolism held at The Royal Melbourne Zoo in February 1995 as a satellite of the XIIth International Conference on Calcium Regulating Hormones (ICCRH). The meeting attracted almost 100 delegates from around the world. Comparative workshops associated with the ICCRH meetings were held regularly up until the last one in Nice in 1986. Since then we have had meetings on various aspects of the comparative endocrinology of calcium regulation associated with other conferences. In 1993, Gert Flik, Neil Hazon and Tim Cheek organised a very successful calcium symposium at the Joint Conference of the European Society for Comparative Physiology and the Society for Experimental Biology in Canterbury, UK. For the first time the scope was widened to include invertebrates, notably crustacean calcium metabolism, as well as delving deeply into the very active area of intracellular calcium homeostasis. At that meeting Gert Flik, Sjeord Wendelaar Bonga and Chris Dacke agreed that it would be timely to revive the comparative calcium regulation symposium in association with the ICCRH.

The scope of the meeting was kept as wide as possible: we had a feast of papers ranging from the presence and role of PTH-like peptides in molluscs to the mammalian placental work which is a strong feature of the various groups based in Melbourne. In addition we had fishes, amphibia, and birds on the menu and even an osteoporotic dinosaur vertebra. Overview papers were provided by five speakers. The first dealt with the presence and possible neuronal function of calcitrophic peptides in invertebrates and vertebrate species. Others reviewed the regulation of mineral balance in invertebrates, the special problems of fish calcium regulation, avian bone and calcium regulation, and mammalian calcium regulation with special reference to PTHrP and placental calcium transport. Carol Gay, who gave the avian overview, found herself at times in stiff competition with the gibbons across the lake, who also wanted to make a loud contribution. The overview talks were interspersed by papers on subjects ranging from the comparative biology of calcium receptors to the biology of molecular cloning of amphibian PTH and PTHrP receptors. We also learned about the latest research on teleocalcin and on vitamin D in teleost fishes, in the mammalian fetus and in subterranean mole rats (which regulate calcium quite happily without the use of vitamin D).

The first afternoon's session was followed by a poster session and a conference barbecue. The following morning we moved from invertebrates, fishes and amphibia to birds and mammals. Finally David Fraser gave an excellent overview summary of the whole meeting. We were greatly privileged to have as chairpersons for the various sessions Jack Martin, Pat Ingleton, Felix Bronner, Shmuel Hurwitz and Hank Kronenburg.

The present meeting stressed the viability, importance and added value of a comparative approach to the study of the endocrinology of calcium metabolism. Its success resulted from the input of a blend of researchers contributing from a phylogenetic perspective. Molecular biology has assumed significant proportions and provided unprecedented tools to elucidate the structure of calcitrophic hormones, their receptors, Ca^{2+} receptors, phylogenetic trees; enterprises once tedious are now readily attainable goals. PTH, once confined to terrestrial vertebrates, is read from a gene that encodes a PTH-related peptide and, intriguingly, this protein may eventually prove to be an ancestral calcitrope found in even the more primitive classes of vertebrates and invertebrates. At the same time, little is known of mineralisation processes in lower vertebrates; here, studies on fish as versatile animal models may yield important new insights. The many questions raised and new ideas generated by this workshop require a 'hands on' approach. The contributors left wonderful Melbourne with new ideas and friends.

The organisers were Chris Dacke, Gert Flik, Ivan Caple and Janine Danks, who was largely responsible for the local arrangements and the delightful venue of the lakeside conference centre. Funds were generously provided by the University of Melbourne Veterinary School and The Beckman Instrument Co. The meeting was greatly enhanced by the enthusiasm of the delegates and their lively contributions to the discussion following each paper. The editors are grateful to the individual authors who provided manuscripts with great alacrity.

We are already planning the next satellite meeting on the Comparative Endocrinology of Calcium Regulation in conjunction with the ICCRH in San Francisco in 1998.

Chris Dacke, Janine Danks, Ivan Caple and Gert Flik
July 1995

Contents

Part One

Invertebrates and lower invertebrates

The Comparative Endocrinology of Calcium Regulation
Eds C Dacke, J Danks, I Caple & G Flik, pp 3-15
Journal of Endocrinology Ltd, Bristol (1996)

Parathyroid hormone and neural calcium regulation: phylogenetic perspectives

S Harvey and R Wang
Department of Physiology, University of Alberta, Edmonton T6G 2H7,
Canada (R Wang is presently at Department of Physiology, University of
Montreal, Montreal, Quebec H3C 3J7)

Introduction

Parathyroid glands were first described in 1880 by Sandstrom (67), and a hypercalcemic factor, parathyroid hormone (PTH) was subsequently isolated from them (14). It has since been established that these glands phylogenetically appear in the tetrapods, coincident with the transition from aquatic to terrestrial environments and with the evolution of bone as a labile calcium (Ca^{2+}) store. Parathyroid hormone or PTH-like peptides are, however, present in more primitive vertebrates and invertebrates. In these species lacking parathyroid glands, PTH-like peptides are present in the central (CNS) or peripheral (PNS) nervous systems. Parathyroidal PTH may thus have evolved from an ancestral neuropeptide, especially as ectopic transcription of the PTH gene occurs in the rat brain and pituitary gland (19, 54). Neurological actions of PTH have also now been demonstrated in vertebrate and invertebrate species and these actions have been shown to be Ca^{2+}-dependent. (17,18, 34, 45, 78, 79). It is possible that the ancestral function of PTH may have been neural Ca^{2+} regulation rather than whole-body Ca^{2+} homeostasis. The evolution of PTH expression and Ca^{2+} regulation are thus the focus of this brief review.

Neural PTH: ectodermal origin?

Parathyroid glands have been considered to be of entodermal origin and to arise embryologically from derivatives of the ventral portion of the third and fourth pharangeal pouches. The PTH-secreting chief cells nevertheless share some ultrastructural and cytochemical characteristics of cells belonging to the APUD series, that are derived from neuroectoderm, although they lack the ability to produce and store amines (63). The entodermal pouches are, however, joined by ectodermal placodes that lie between them and ectodermal tissue may contribute to parathyroid gland development.

Neural PTH: occurrence in tetrapods

The occurrence of PTH-like peptides in the CNS is now well-established (see (31) for a review). Extracts of mammalian (sheep, rat, rabbit, mouse, hamster), avian (quail, chicken, duck), reptilian (turtle, garter snake) and amphibian (toad, bullfrog, mud puppy) brains and pituitary glands have been shown to contain PTH immunoreactivity (IR) that coeluted with PTH1-84 after HPLC separation of the extracts (6, 33, 59). The presence of a PTH-like peptide in the brain was also indicated by Western blotting (54), which showed that neural PTH-IR was associated with proteins of comparable size to those of the parathyroid gland. This immunoreactivity is widely distributed throughout the brain but is most abundant in hypothalamic regions (60). Indeed, cell bodies reacting with PTH antisera specific for the 48-64 region of bovine PTH(1-84) are located in preoptic nuclei of the mouse and bullfrog brain (60). Immunoreactive fibers from these cell bodies can be traced to the median eminence, where they terminate around pituitary portal blood vessels. IR-PTH has also been detected in human, rat and rabbit cerebrospinal fluid (CSF), although at concentrations 20-70% less than those in peripheral plasma (5, 41).

Neural PTH: central or peripheral origin?
The blood-brain barrier has been considered by many authors to be impermeable to PTH in systemic circulation. The poor correlation between the concentration of PTH in plasma and CSF of many human patients supports this view (31, 38), although plasma: CSF PTH levels are directly correlated in some individuals (11, 23). Indeed, the implantation of parathyroidal tissue into the brain of parathyroidectomized rats maintains the concentration of PTH in peripheral plasma within the normal range (85). Thus, while the presence of PTH-like peptides in the hypothalamo-pituitary axis may reflect the local expression of a PTH-like gene or ectopic transcription of the PTH gene, it could reflect uptake from systemic circulation, especially as the pituitary gland and median eminence lie outside the blood-brain barrier.

Neural PTH: ectopic transcription of the parathyroidal PTH gene?
Expression of a PTH-like gene in neural tissue was demonstrated by Northern blotting and *in situ* hybridization of rat brain tissues with rat parathyroidal PTH gene probes (19). An RNA moiety of identical size to PTH mRNA was detected in hypothalamic tissue and localized in hypothalamic nuclei. The nucleotide sequence of this mRNA moiety was subsequently shown by RT-PCR to be identical in the peptide coding region to that from the parathyroid gland (54).

Parathyroidal PTH: ectopic transcription of the neural PTH gene?
Parathyroid glands are not ancestral sites of PTH synthesis, since PTH-like peptides have been detected in extracts of fish (eel, cod, hake, goldfish, trout, salmon) pituitary glands (20, 33, 42, 43, 62), which Pearse and Takor (63) consider to be of ectodermal origin. As in tetrapods, this PTH-IR in fish pituitary glands is comparable to PTH1-84 in molecular size. This immunoreactivity is largely confined to the neurohypophysis, in the terminals of nerve fibers emanating from the preoptic area of the hypothalamus (42). In more primitive vertebrates PTH-IR has also been located in the preoptic area of the hagfish brain and in the neurohypophysis of this agnathan (20). PTH-like IR has also been detected in whole body extracts of the protochordate *Amphioxus* (unpublished observations) and in neural ganglia of the mollusk *Lymnea*

stagnalis (83). Neural PTH, unlike parathyroidal PTH, therefore has an ancient phylogenetic history that can be traced as far back as the invertebrates. It is therefore possible that PTH gene expression in the parathyroid gland reflects the ectopic transcription of the neural PTH gene.

Neural PTH: a peptide family?
Although a single PTH gene is transcribed in the parathyroid gland, a PTH-like gene family may be expressed in the hypothalamo-pituitary axis. In addition to PTH, the gene for PTH-related protein (PTHrP) is widely expressed within the brain (80) in which a number of PTHrP-IR proteins have been identified (55). Heterogeneous PTH-like peptides may also be present in neural tissues, since multiple cDNAs are generated by RT-PCR of hypothalamic RNA after amplification with oligonucleotide primers for parathyroidal PTH cDNA (54). The absence of this heterogeneity in parthyroidal cDNA suggests differential (tissue-specific) transcription of the PTH gene in the brain.

Neural PTH: evolution of the PTH gene
The PTH gene was recently considered by Mallette (51) to have arisen from the duplication and reduplication of a primitive gene. PreproPTH contains a fourfold internal homology that may represent four copies (spanning sequences -30 to -6, -3 to $+22$, $+23$ to $+47$ and $+50$ to $+74$) of an ancient gene that coded for a 25 amino acid peptide. This gene may have been related to proopiomelanocortin (POMC), since preproPTH has significant homology with sequences of adrenocorticotropin (ACTH) and melanocyte stimulating hormone (MSH) that are coded by POMC. It has therefore been proposed that PTH and POMC arose from a common precursor. The primitive PTH gene may thus have appeared early during evolution, since IR-ACTH has been determined in bacteria (24) and in the unicellular flagellate, *Tetrahymena pyriformis* (46). Moreover, although the POMC gene is widely expressed in the tissues of higher vertebrates, it would appear to have been initially expressed in the neural tissues of multicellular invertebrates (2, 3, 77, 84). Like PTH, POMC peptides are also located in the sensory ganglia of mollusks (8), in which ACTH has been shown to enhance the chemosensitivity of neural membranes and induce electrical activity (75). Interestingly, POMC-like peptides are also present in the immune cells of gastropods (56), in which they have roles in modulating phagocytosis and chemotaxis, and in the release of biogenic amines (57). PTH has also been found to affect lymphocyte proliferation and function (4, 47) and has been shown to be ectopically transcribed in immune cells of the mammalian immune system (22, 30).

The primitive gene coding for a PTH-like peptide may have also been subjected to chromosomal duplication, giving rise to the PTHrP gene, which is located on chromosome 11 in a similar position to the PTH gene on chromosome 12 (51). Eight of the first 13 amino acids of these peptides and the last two amino acids of their prohormone sequences (16-18) are identical. Moreover, although the rest of their amino acid sequences differ markedly, significant identity still exists between them in the base nucleotides present in codon positions 1 and 2 of the codons present in three

5

of the four homologous regions of preproPTH. There is, however, no evidence that preproPTHrP now preserves the fourfold homology found within preproPTH. The chromosomal duplication that probably gave rise to PTH and PTHrP may therefore have occurred soon after at least three of the four homologous regions in preproPTH had been formed. The internal homology that may have been originally present within preproPTHrP may thus have been subjected to less evolutionary constraint than those in the N-terminal receptor binding domain and subsequently lost.

Neural PTH and calcitropic hormones
Many hormones participate in whole-body Ca^{2+} homeostasis in higher vertebrates, but it is largely maintained through the opposing hypercalcemic and hypocalcemic actions of PTH and calcitonin respectively (58, 61). Extracellular Ca^{2+} regulation is, however, unlikely to have been the ancestral role of either hormone. Indeed, IR-calcitonin is present in unicellular bacteria (65) and protozoans (16), prior to the evolutionary development of extracellular communication. Moreover, prior to the evolutionary development of thyroid glands and ultimobranchial bodies in vertebrates, calcitonin is present in the central and peripheral nervous systems of invertebrates (26, 27, 68, 69), as it remains in vertebrates (64, 69). Furthermore, while calcitonin may regulate extracellular $[Ca^{2+}]$ concentrations in multicellular invertebrates, by modulating intestinal Ca^{2+} uptake or exoskeletal Ca^{2+} deposition (10, 26), it acts as a neurotrophic factor (29) or neuromodulator (70) in phylogenetically primitive animals.

In addition to PTH and calcitonin, extracellular Ca^{2+} homeostasis in vertebrates is regulated by vitamin D (39), growth hormone (GH) and prolactin (81). These hormones have similarly been located in invertebrate nervous systems (7, 35, 71, 74) and are also found in the brain and pituitary glands of evolutionary advanced vertebrates (28, 35, 53). It is therefore not surprising that these calcitropic hormones have neurological actions (35, 66, 73). Recent studies have also demonstrated roles for PTH in neural function (1, 13, 17, 18, 34). The nervous system would thus appear to be the evolutionary origin of many of the hormones now responsible for extracellular or whole-body Ca^{2+} homeostasis. The evolution of this homeostatic role may have arisen from ancestral calcemic actions of these factors on neural cells, since neurological activity is Ca^{2+}-dependent.

The possibility that whole-body Ca^{2+} homeostasis was ancestrally regulated by neural factors is supported by the dominant role of the pituitary gland in fish Ca^{2+} homeostasis (81). It is only in the terrestrial vertebrates that PTH and calcitonin predominate in the endocrine control of calcium and phosphate metabolism. Indeed, it is generally considered that the pituitary gland is the main source of hypercalcemic hormones in fish (81). Fish become hypocalcemic after the removal of the pituitary gland, and this can be corrected by injections of pituitary homogenates or by pituitary transplantation. This may be due to the presence of a factor tentatively localized in the pars intermedia called hypercalcin that has cross-reactivity with several antisera against mammalian PTH (33, 62, 81), although mammalian PTH may not be hypercalcemic in fish (82). Prolactin can also overcome the hypocalcemic effect of hypophysectomy and is a potent hypercalcemic factor in fish, as in higher vertebrates

(58, 61, 81). ACTH is also hypercalcemic, although probably as a result of its induction of cortisol. Growth hormone also has hypercalcemic activity in higher vertebrates (81).

Although fish have ultimobranchial glands that secrete calcitonin, this hormone has little, if any, hypocalcemic action in fish (15). Hypocalcemia is, instead, mediated through the actions of stanniocalcin, derived from the corpuscles of Stannius (15, 37) which also has some PTH-like immunoreactivity (33, 48). Interestingly, this hormone has also been immunologically detected in the brain and pituitary gland of the platyfish, *Xiphophorus maculatus* (20). Stanniocalcin immunoreactivity, distinct from PTH-IR or PTHrP-IR, is mainly located in the pars intermedia and in nerve fibers in the pars distalis, although no adenohypophyseal cells are positive. In the brain, IR cell bodies are found in the nucleus preopticus, similar to the distribution of IR-PTH in other species. The occurrence of this hypocalcemic peptide in the neural ganglia of snails (83) also suggests that both hypocalcemic and hypercalcemic hormones were originally expressed in neural tissues.

Neural PTH: neural roles
Numerous neurological and psychiatric disturbances occur in hyperparathyroid disease states (1, 31). Intraventricular injections of PTH in the rat brain also modulate centrally regulated physiological functions (13). The possibility that these reflect direct actions of PTH on neural membranes is supported by the presence of PTH receptor mRNA in the rat brain (36, 76) and the occurrence of PTH-binding sites on rat (32) and snail (44) neural membranes. Rat brain synaptosomes are, moreover, responsive to PTH stimulation *in vitro* (18, 49), resulting in increased release of γ-aminobutyric acid (49). The release of glutamyldopamine from isolated snail neurons is also directly increased by PTH (78), as is the turnover of dopamine in the rat hypothalamus (34). Neuromodulatory actions of PTH are also indicated by its generation of inward Ca^{2+} currents in the isolated neurons of the snail, *Helisoma trivolis* (78, 79) and *Helix potamia* (45). The entry of Ca^{2+} into neural tissue is thus likely to promote neurotransmitter release.

Neural PTH and neural Ca^{2+} regulation
Although PTH can directly modulate neuronal function, many of the neurological disturbances associated with hyperparathyroidism are thought to be secondary consequences of increased extracellular Ca^{2+} concentrations, $[Ca^{2+}]_e$. The entry of Ca^{2+} into neural tissue is stimulated by elevated $[Ca^{2+}]_e$ and this modulates the electrical properties of neuronal membranes and promotes the exocytotic discharge of secretory vesicles. The entry of Ca^{2+} into neural tissue is, however, also directly stimulated by PTH.

The entry of Ca^{2+} into cells occurs through voltage-dependent Na^+ channels, voltage-dependent Ca^{2+} channels, ligand- or second messenger-dependent Ca^{2+} channels and, in neural tissue, by activation of a Na^+/Ca^{2+} exchanger. Ca^{2+} extrusion is mediated by Ca^{2+} ATPase and Na^+/K^+ pumps and by the Na^+/Ca^{2+} exchanger (9, 17). These mechanisms serve to extrude Ca^{2+} against a Ca^{2+} gradient in non-excitable cells, but facilitate Ca^{2+} entry into nervous tissue. In response to PTH stimulation, ATP-dependent Ca^{2+} transport and Na^+/Ca^{2+} exchange are rapidly (within 5 s) activated in rat brain synaptosomes (17).

Although both Ca^{2+} influx and Ca^{2+} efflux are stimulated by PTH, the rate of influx exceeds the rate of efflux, increasing Ca^{2+} concentrations $[Ca^{2+}]_i$. The extrusion of Ca^{2+} from rat brain synaptosomes is suppressed following chronic PTH stimulation, as a result of the inhibitory effect of high $[Ca^{2+}]_i$ on mitochondrial oxidation and ATP production, which leads to an impairment of the Ca^{2+} ATPase and Na^+/K^+ATPase pumps (17). This synaptosomal rise in $[Ca^{2+}]_i$ might also result from the activation of phosphoinositol metabolism and the formation of diacylglycerol (DAG), inositol triphosphate (IP_3) and inositol 1,3,4,5-tetrakisphosphate (IP_4) (9, 45). IP_4 by itself and/or with IP_3 or DAG may change membrane permeability of the cell to facilitate extracellular Ca^{2+} influx and may stimulate the release of microsomal Ca^{2+} stores (45).

In addition to stimulating Ca^{2+} transport into synaptosomes, PTH also induces a biphasic increase in Ca^{2+} entry into isolated snail neurons. A rapid influx of Ca^{2+} accompanies a transient increase in Ca^{2+} current and is followed by a slow and sustained increase in $[Ca^{2+}]_i$ that may last for more than 15 min (78, 79) or 1 h (45). This induction of Ca^{2+} transport occurs through a mechanism that is additive with KCl-induced $[Ca^{2+}]_i$ and is induced in a dose-related way by PTH1-34 or PTH1-84. This increase in $[Ca^{2+}]_i$ is dependent upon the influx of extracellular Ca^{2+} through N-like voltage-dependent Ca^{2+} channels that are blocked by lanthanum and ω-conotoxin (79). The influx of Ca^{2+} into snail neurons is not, however, mediated through voltage-dependent Na^+ or K^+ channels, nor via a Na^+/Ca^{2+} exchanger. The influx of Ca^{2+} into these neurons is dependent upon ATP, as in synaptosomes (45). The uptake of Ca^{2+} is thus associated with an energy-dependent process that is likely to result in the phosphorylation of membrane channel proteins.

Intracellular metabolic processes therefore contribute to the PTH-induced increase in $[Ca^{2+}]_i$ and these are thus likely to reflect the stimulation of intracellular signal transduction systems following the binding of PTH to membrane receptors. Indeed, the effect of PTH on N-like Ca^{2+} channel currents in *Helisoma* neurons is dependent upon the activation of a G protein (78). This protein is insensitive to pertussis toxin and is unlikely to activate adenylate cyclase-dependent protein kinase activity, as the action of PTH on $[Ca^{2+}]_i$ is not mimicked by dibutyryl cAMP nor blocked by inhibition of cyclic adenosine monophosphate (cAMP)-dependent protein kinase activity. Activation of the adenylate cyclase system did, however, potentiate $[Ca^{2+}]_i$ entry in some *Helix* neurons, but this action was additive with PTH and had a dissimilar time course of action (45). Similarly, although cAMP analogs and adenylate cyclase activity increase Ca^{2+} transport into rat brain synaptosomes, PTH-stimulated Ca^{2+} influx into rat brain synaptosomes is independent of cAMP activation (18). The entry of Ca^{2+} into snail neurons may, however, involve the activation of a protein kinase C (PKC) pathway, since the action of PTH is blocked by PKC inhibition and mimicked by PKC-activating phorbol esters (45). Moreover, the Ca^{2+}-dependent activation of PKC is likely to account for the biphasic pattern of $[Ca^{2+}]_i$, the early phase being PKC-independent and the later phase being PKC-dependent and possibly accompanied by DAG and IP_3 formation and the mobilization of intracellular Ca^{2+} stores.

PTH stimulation of neural $[Ca^{2+}]_i$: phylogenetic perspectives

Neural communication is dependent upon the influx of extracellular Ca^{2+} to stimulate neurotransmitter release through the exocytotic discharge of secretory vesicles. The rapidity of neural communication is dependent upon the rapid entry of extracellular Ca^{2+} into neural tissues. Fluctuations in extracellular $[Ca^{2+}]_e$ as a result of whole-body Ca^{2+} homeostasis would occur too slowly for efficient neural communication and neural firing would lack specificity. The rapid modulation of $[Ca^{2+}]_i$ in neural cells may thus have been an ancestral role of PTH and other neurohormones (e.g. calcitonin, vitamin D) now associated with whole-body Ca^{2+} homeostasis. In more primitive animals, the epithelial membranes of the kidney, skin and gut are thought to have facilitated Ca^{2+} movement between the internal and external (aquatic) Ca^{2+}-rich environment. The possibility that PTH may regulate extracellular Ca^{2+} homeostasis in invertebrates has, however, yet to be rigorously examined. The exoskeleton of some crustaceans is, nevertheless, mobilized by PTH to increase hemolymph Ca^{2+} concentrations (50), contrary to the effects of calcitonin (72).

PTH stimulation of neural $[Ca^{2+}]_i$: relationship with extracellular $[Ca^{2+}]_e$

The neurological actions of PTH are primarily induced as a result of the increased entry of extracellular Ca^{2+} into neural tissues. It has, however, also been suggested that PTH increases $[Ca^{2+}]_i$ as a result of the mobilization of intracellular Ca^{2+} from microsomal stores. Similar mechanisms of PTH action are also utilized in non-neural target tissues to stimulate the intracellular pathways that result in bone resorption, renal Ca^{2+} retention and elevated concentrations of extracellular Ca^{2+} (12, 21, 40). It is therefore possible that PTH actions in these 'traditional' target tissues have evolved from ancestral actions of PTH on neural Ca^{2+} regulation. Neural Ca^{2+} regulation and extracellular Ca^{2+} regulation are also functionally related. This was recently demonstrated by Matsui *et al.* (52) by the hypercalcemia induced in rats following intraventricular injections of PTH. Actions of PTH within the brain are thus likely to stimulate neural and/or neuroendocrine pathways that promote bone resorption or renal Ca^{2+} retention or block central pathways that induce hypocalcemia. This relationship between central and peripheral calcemic actions of PTH is also mirrored by calcitonin, which has been found to induce hypocalcemia in peripheral plasma following its interaction with CNS receptors (25).

PTH regulation of neural Ca^{2+}: phylogenetic summary

The occurrence of PTH-like peptides in the nervous systems of species lacking parathyroid glands and the ectodermal origin of parathyroid glands suggests parathyroidal PTH evolved from a neuropeptide expressed in ancestrally primitive neurons. Calcitonin, hypocalcin and other Ca^{2+} regulating hormones are also likely to have evolved in the nervous systems of phylogenetically ancient species.

The hypercalcemic actions of PTH in tetrapods probably evolved from the neuromodulatory actions of PTH within primitive neurons, mediated through increased extracellular Ca^{2+} influx, increased intracellular Ca^{2+} mobilization and increased $[Ca^{2+}]_i$.

Acknowledgements
This work was supported, in part, by grants from the Medical Research Council and the Natural Sciences and Engineering Research Councils of Canada.

References

1   Akmal M, Goldstein DA, Mutani S & Massry SG (1984) Role of uraemia, brain calcium and parathyroid hormone on changes in electroencephalogram in chronic renal failure. *American Journal of Physiology* **246** F575-F579.

2   Al-Yousuf S (1990) Neuropeptides in annelids. *Progress in Comparative Endocrinology* pp 232-241. Eds A Epple, CG Scanes & MT Stetson. New York: Wiley-Liss.

3   Aros B, Wenger T, Vigh B & Vigh-Teichman I (1980) Immunohistochemical localization of substance P and ACTH-like activity in the central nervous system of the earthworm, *Lumbricus terrestris L. Acta Histochemica* **66** 262-268.

4   Atkinson MJ, Hesch RD, Cade C, Wadwah M & Perris AD (1987) Parathyroid hormone stimulation of mitosis in rat thymic lymphocytes is independent of cyclic AM. *Journal of Bone and Mineral Research* **2** 303-309.

5   Balabanova S, Tollner U, Richter HP, Pohlandt F, Gaedicke G & Teller W M (1984) Immunoreactive parathyroid hormone, calcium and magnesium in human cerebrospinal fluid. *Acta Endocrinologica* **106** 227-233.

6   Balabanova S, King O, Teller WN & Reinhardt G (1985) Distribution and concentration of immunoreactive parathyroid hormone in brain and pituitary of sheep. *Klinische Wochenschrift* **63** 419-422.

7   Bidmon HJ & Stumpf WE (1991) Uptake, distribution and binding of vertebrate and invertebrate steroid hormones and time-dependence of porasterone A binding in *Calliphora vicina. Histochemistry* **96** 419-434.

8   Boer HH, Schott LP, Roubos EW, ter Maat A, Lodder JC, Recchelt D & Swabb DF (1979) ACTH-like immunoreactivity in two electronically coupled giant neurons in the pond snail, *Lymnaea stagnalis. Cell and Tissue Research* **202** 231-240.

9   Brown EM (1994) Homeostatic mechanisms regulating extracellular and intracellular calcium metabolism. In *The Parathyroids*, pp 15-54. Eds JP Bilezikian, MA Levine & R Marcus. New York: Wiley-Liss.

10   Cameron JN & Thomas P (1992) Calcitonin-like immunoreactivity in the blue crab: tissue distribution, variation in the molt cycle and partial characterization. *Journal of Experimental Zoology* **262** 279-286.

11   Care AB & Bell NH (1986) Evidence that parathyroid hormone crosses the blood brain barrier. *Proceedings of the IXth International Conference on Calcium Regulating Hormones and Bone Metabolism,* Nice p 181.

12   Civitelli R, Fujimori A, Bernier SM, Warlow PM, Goltzman D, Hruska KA & Avioli LV (1992) Heterogeneous intracellular free calcium responses to parathyroid hormone correlate with morphology and receptor distribution in osteogenic sarcoma cells. *Endocrinology* **130** 2392-2400.

13   Clementi G, Drago F, Amico-Roxas M, Prato A, Raisarda E, Nicoletti F, Rodolico G & Scapagnini U (1985) Central actions of calcitonin and parathyroid hormone. In *Calcitonin: Chemistry, Physiology and Clinical Aspects*, pp 279-286. Ed A Pecile. Amsterdam: Elsevier Science Publishers BV.

14   Collipp JB & Clark EP (1925) Further studies of the physiological action of a parathyroid hormone. *Journal of Biological Chemistry* **66** 133-138.

15   Copp DH & Kline LW (1989) Calcitonin. In *Vertebrate Endocrinology: Fundamentals and Biomedical Implications,* vol 3, pp 79-104. Eds PKT Pang & MP Schreibman. New York: Academic Press.

16   Deftos LJ, Le Roith D, Shiloach J & Roth J (1985) Salmon calcitonin-like immunoreactivity in extracts of *Tetrahymena pyriformis. Hormones and Metabolic Research* **17** 82-85.

17   Fraser CL & Arieff AI (1988) PTH increases Ca++ transport in rat brain synaptosomes in uremia. *New Actions of Parathyroid Hormone*, pp 317-328. Eds SG Massry & F Fujita. New York: Plenum Press.

18   Fraser CL, Sarnacki P & Budayr A (1988) Evidence that parathyroid hormone-mediated calcium transport in rat brain synaptosomes is independent of cyclic adenosine monophosphate. *Journal of Clinical Investigation* **81** 982-988.

19   Fraser RA, Kronenberg HM, Pang PKT & Harvey S (1990) Parathyroid hormone mRNA in the rat hypothalamus. *Endocrinology* **127** 2517-2522.

20   Fraser RA, Kaneko T, Pang PKT & Harvey S (1991) Hypo- and hypercalcemic peptides in fish pituitary glands. *American Journal of Physiology* **260** R622-R626.

21   Fukayama S & Tashjian AH Jr (1990) Stimulation by parathyroid hormone of $45Ca^{2+}$ uptake in osteoblast-like cells: possible involvement of alkaline phosphatase. *Endocrinology* **126** 1941-1949.

22   Fukumoto S, Matsumoto T, Watanabe T, Takahashi H, Miyoshi I & Ogata E (1989) Secretion of PTH-like activity from human cell lymphotropic virus type 1-infected lymphocytes. *Cancer Research* **49** 3849-3852.

23   Gennari C (1988) Parathyroid hormone and pain. In *New Actions of Parathyroid Hormone*, pp 335-344. Eds SG Massry & T Fujita. New York: Plenum Press.

24   Ghione M & Dell'Orto P (1983) Human polypeptidic hormone-like substances in microorganisms. *Microbiologica* **6** 315-326.

25   Goltzman D & Tannenbaum GS (1987) Induction of hypocalcemia by intraventricular injection of calcitonin: evidence for control of blood calcium by the nervous system. *Brain Research* **416** 1-6.

26   Graf F, Fouchereau-Peron M, Van-Wormhandt A & Meyran JC (1989) Variations of calcitonin-like immunoreactivity in the crustacean, *Orchestia cavimara*, during a molt cycle. *General and Comparative Endocrinology* **73** 80-84.

27   Graf F, Morel G & Meyran JC (1992) Immunocytological localization of endogenous calcitonin-like molecules in the crustacean, *Orchestia. Histochemistry* **97** 147-154.

28   Grifford B, Deray A, Jacquemard C, Fellmann D & Bugno C (1994) Prolactin immunoreactive neurons of the rat lateral hypothalamus: immunocytochemical and ultrastructural studies. *Brain Research* **635** 179-186.

29   Grimm-Jorgensen Y (1987) Somatostatin and calcitonin stimulate neurite regeneration of molluscan neurons *in vitro. Brain Research* **403** 121-126.

30   Handt O, Reis A & Schidtke J (1992) Ectopic transcription of the parathyroid hormone gene in lymphocytes, lymphoblastoid cells and tumor tissue. *Journal of Endocrinology* **135** 249-256.

31   Harvey S & Fraser RA (1993) Parathyroid hormone: neural and neuroendocrine perspectives. *Journal of Endocrinology* 139 353-361.

32   Harvey S & Hayer S (1993) Parathyroid hormone binding sites in the brain. *Peptides* **14** 1187-1191.

33   Harvey S, Zeng Y-Y & Pang PKT (1987) Parathyroid hormone-like immunoreactivity in fish plasma and tissues. *General and Comparative Endocrinology* **68** 136-146.

34   Harvey S, Hayer S & Sloley BD (1993) Parathyroid hormone-induced dopamine turnover in the rat medial basal hypothalamus. *Peptides* **14** 269-274.

35   Harvey S, Hull KL & Fraser RA (1993) Growth hormone: Neurocrine and neuroendocrine perspectives. *Growth Regulation* **3** 161-171.

36   Hashimoto H, Aino H, Ogawa N, Nagata S & Baba A (1994) Identification and characterization of parathyroid hormone/parathyroid hormone-related peptide receptor in cultured astrocytes. *Biochemical and Biophysical Research Communications* **200** 1042-1048.

37   Hirano T (1989) The corpuscles of Stannius. In *Vertebrate Endocrinology: Fundamentals and Biomedical Implications* pp 139-170. Eds P K T Pang & M P Schreibman. New York: Academic Press.

38   Hironaka T, Morimoto S, Fukuo K, Koh E, Imanaka S, Sumida T, Onishi T & Kumahara Y (1987) Immunoreactive parathyroid hormone in the circulation and cerebrospinal fluid from patients with renal failure: possible restriction of parathyroid hormone in the blood-brain barrier. *Bone and Mineral* **2** 487-494.

39   Holick MF (1989) Phylogenetic and evolutionary aspects of vitamin D from phytoplankton to humans. In *Vertebrate Endocrinology: Fundamentals and Biomedical Implications*, pp 7-44. Eds PKT Pang & MP Schreibman. New York: Academic Press.

40   Hruska KA, Goligorsky M, Scoble J, Tsutsumi M, Westbrook S & Moskowitz D (1986) Effects of parathyroid hormone on cytosolic calcium in renal proximal tubular primary cultures. *American Journal of Physiology* **251** F188-F198.

41   Joborn C, Hetta J, Niklasson F, Rasstad J, Wide L, Agren H, Akerstrom G & Ljunghall S (1991) Cerebrospinal fluid, calcium, parathyroid hormone and monoamine and purine metabolites and the blood-brain barrier function in primary hyperparathyroidism. *Psychoneuroendocrinology* **16** 311-322.

42   Kaneko T & Pang PKT (1987) Immunocytochemical detection of parathyroid hormone-like substance in the goldfish brain and pituitary gland. *General and Comparative Endocrinology* **68** 147-152.

43   Kaneko T, Harvey S, Kline LW & Pang PKT (1989) Localization of calcium regulatory hormones in fish. *Fish Physiology and Biochemistry* **7** 337-342.

44   Khudaverdyan DN, Ter-Markosyan AS & Sargasyan AR (1991) On the mechanism of parathyroid hormone effect on functional activity of nerve cells. *Neurochemistry* **8** 210-215.

45   Kostyuk PG, Lukyanetz EA & Ter-Markosyan AS (1992) Parathyroid hormone enhances calcium current in snail neurones: stimulation of the effect by phorbol esters. *Plügers Archives* **420** 146-152.

46   Le Roith D, Liotta AS, Roth J, Shiloach ME, Lewis ME, Pert CB & Krieger DT (1982) ACTH and endorphin-like materials are native to unicellular organisms. *Proceedings of the National Academy of Sciences of the USA* **79** 2086-2090.

47   Lewin E, Ladefoged J, Brandi L & Olgaard K (1993) Parathyroid hormone dependent T cell proliferation in uremic rats. *Kidney International* **44** 379-384.

48   Lopez E, Tisserand-Jochem EM, Eyquen A, Milet C, Hillyard E, Lallier F, Vidal B & MacIntyre I (1984) Immunocytochemical detection in eel corpuscles of Stannius of a mammalian parathyroid hormone-like hormone. *General and Comparative Endocrinology* **53** 28-36.

49   Lutsenko VR, Ter-Markosyan AS, Khlebrikova NN & Kudaverdyan DJ (1987) Effect of parathyroid hormone on the transport of $^{45}Ca^{2+}$ and ^{3}H-GABA in nerve endings isolated from the rat cerebral cortex. *Byulleten Eksperimentalnoi Brologii i Meditsiny* **104** 146-149.

50   McWhinnie DJ, cited by Pandey AJ (58).

51   Mallette LE (1994) Parathyroid hormone and parathyroid hormone-related protein as polyhormones. Evolutionary aspects and non-classical actions. In *The Parathyroids*, pp 171-184. Eds JP Bilezikian, MA Levine & R Marcus. New York: Raven Press.

52   Matsui H, Aoo S, Ma J & Hori T (1995) Central actions of parathyroid hormone on blood calcium and hypothalamic neuronal activity in the rat. *American Journal of Physiology* **268** R21-R27.

53   Neveu I, Naveilhan P, Menaa C, Wion D, Braket P & Garabedian M (1994) Synthesis of 1-25 dihydroxyvitamin D3 by rat brain macrophages *in vitro*. *Journal of Neuroscience Research* **38** 214-220.

54   Nutley MT, Parimi SA & Harvey S (1995) Sequence analysis of hypothalamic parathyroid hormone messenger ribonucleic acid. *Endocrinology* **136** 5600-5607.

55   Orloff JB, Wu TL & Stewart AF (1989) Parathyroid hormone-like proteins: biochemical responses and receptor interaction. *Endocrine Reviews* **610** 476-495.

56   Ottaviani E & Cossarizza A (1990) Immunocytochemical evidence of vertebrate bioactive peptide-like molecules in the immuno cell types of the freshwater snail, *Planobarius corneus* (L.) (Gastropoda, Pulmonata). *FEBS Letters* **267** 250-252.

57   Ottaviani E, Petraglia F, Montagnani G, Cossarizza A, Monti D & Franceschi C (1990) Presence of ACTH and b-endorphin immunoreactive molecules in the freshwater snail, *Planorbarius corneus* (L.) (Gastropoda, Pulmonata) and their possible role in phagocytosis. *Regulatory Peptides* **27** 1-9.

58   Pandey AK (1992) Endocrinology of calcium metabolism in amphibians, with emphasis on the evolution of hypercalcemic regulation in tetrapods. *Biological Structures and Morphogenesis* **4** 102-126.

59   Pang PKT, Harvey S, Fraser RA & Kaneko T (1988) Parathyroid hormone-like immunoreactivity in brains of tetrapod vertebrates. *American Journal of Physiology* **255** R635-R642.

60   Pang PKT, Kaneko T & Harvey S (1988) Immunocytochemical distribution of PTH-immunoreactivity in vertebrate brains. *American Journal of Physiology* **255** R643-R647.

61   Pang PKT, Pang RK & Wendelaar Bonga SE (1989) Hormones and calcium regulation in vertebrates: an evolutionary and overall consideration. In *Vertebrate Endocrinology: Fundamentals and Biomedical Implications*, vol 3, pp 343-352. Eds PKT Pang & MP Schreibman. New York: Academic Press.

62   Parsons JA, Gray D, Rafferty G & Zanelly JM (1978) Evidence for a hypercalcemic factor in the fish pituitary, immunologically related to mammalian parathyroid hormone. *Endocrinology of Calcium Metabolism*, pp 111-114. Eds DH Copp & RV Talmage. Amsterdam: Excerpta Medica.

63   Pearse AGE & Takor TT (1976) Neuroendocrine embryology and the APUD concept. *Clinical Endocrinology* 5 (Suppl) 2295-2445.

64   Perez-Cano R, Girgis SI & MacIntyre I (1982) Further evidence for calcitonin gene duplication: the identification of two different calcitonins in a fish, a reptile and two mammals. *Acta Endocrinologica* 100 256-261.

65   Perez-Cano R, Murphy PK, Girgis SI, Arnett TR, Blenkharm T & MacIntyre I (1982) Unicellular organisms contain a molecule resembling human calcitonin. *The Endocrine Society Annual Meeting* Abstract 673.

66   Phelps CJ (1994) Pituitary hormones as neurotrophic signals: anamalous hypophysiotropic neuron differentiation in hypopituitary dwarf mice. *Proceedings of the Society for Experimental Biology and Medicine* 206 6-23.

67   Sandstrom I (1938) *On a New Gland in Man and Several Mammals*. New York: The Johns Hopkins Press.

68   Sasayama Y, Katoh C, Kambegawa A & Yoshizawa H (1991) Cells showing immunoreactivity for calcitonin gene-related peptide (CGRP) in the central nervous system of some invertebrates. *General and Comparative Endocrinology* 83 406-414.

69   Sasayama Y, Yoshihara M, Fujimori M & Oguro C (1990) The role of the ultimobranchial gland in calcium metabolism of amphibians and reptiles. *Progress in Clinical and Biological Research* 342 592-597.

70   Sawada M, Ichinose M, Ishihawa S & Sasayama Y (1993) Calcitonin induces a decreased Na+ conductance in identified neurons of *Aplysia*. *Journal of Neuroscience Research* 36 200-208.

71   Schmidt KP, Maier V, Haug C, Pfeiffer EF (1990) Ultrastructural localization of prolactin-like antigenic determinants in neurosecretory cells in the brain of the honeybee (*Apis mellifuca*). *Hormone and Metabolic Research* 22 413-417.

72   Sellem E, Graf F & Meyran JC (1989) Some effects of salmon calcitonin on calcium metabolism in the crustacean *Orchestia* during the molt cycle. *Journal of Experimental Zoology* 249 177-181.

73   Stumpf WE (1990) Steroid hormones and the cardiovascular system: direct actions of estradiol, progesterone, testosterone, gluco- and mineralcorticoids and soltirol (vitamin D) on central nervous resgulatory and peripheral hormones. *Experientia* 46 13-25.

74   Swinnen K, Broeck IV, Verhaert P & de Loof A (1990) Immunocytochemical localization of human growth hormone- and prolactin-like antigenic determinants in the insects, *Locusta migratoria* and *Sarcophaga bullata*. *Comparative Biochemistry and Physiology* 95A 373-378.

75   Tsitolovskii LE (1981) Mollusk neuron habituation following extra- and intracellular administration of an ACTH fragment. *Zhurnal Vysshei Mervnoi Deyatelnosti Imeni i P Pavlova* 31 577-586.

76   Urena P, Kong X-F, Abou-Samra A-B, Juppner H, Kronenberg HM, Potts JJ Jr & Segre GV (1993) Parathyroid hormone (PTH)/PTH-related peptide receptor messenger ribonucleic acids are widely distributed in rat tissues. *Endocrinology* 133 617-623.

77   Verhaert P, Ma M & De Loof A (1990) Immunochemistry and comparative insect (neuro) endocrinology. In *Progress in Comparative Endocrinology*, pp 315-322. Eds A Epple, CG Scanes & MT Stetson. New York: Wiley-Liss.

78   Wang R, Pang PKT, Wu L, Shipley A, Karpinski E, Harvey S & Berdan RC (1993) Neural effects of parathyroid hormone: modulation of the calcium channel current and metabolism of monoamines in identified *Helisoma* snail neurons. *Canadian Journal of Physiology and Pharmacology* **71** 582-591.

79   Wang R, Pang PKT, Wu L, Karpinski E, Harvey S & Berdan RC (1994) Enhanced calcium influx by parathyroid hormone in identified *Helisom trivolis* snail neurons. *Cell Calcium* **15** 89-84.

80   Weir EC, Brine ML, Ikeda W, Bustis WJ, Broadus AE & Robbins RJ (1990) Parathyroid hormone-related peptide gene is expressed in the mammalian central nervous system. *Proceedings of the National Academy of Sciences of the USA* **87** 108-112.

81   Wendelaar Bonga SE & Pang PKT (1989) Pituitary hormones. In *Vertebrate Endocrinology: Fundamentals and Biomedical Implications, vol 3, Regulation of Calcium and Phosphate*, pp 105-138. Eds PKT Pang & MP Schreibman. New York: Academic Press.

82   Wendelaar Bonga SE, Pang RK & Pang PKT (1986) Hypocalcemic effects of bovine PTH-(1-34) and Stannius corpuscle homogenates in teleost fish adapted to low calcium water. *Journal Experimental Zoology* **240** 263-267.

83   Wendelaar Bonga SE, Lafeber FPJG, Flik G, Kaneko T & Pang PKT (1989) Immunocytochemical demonstration of a novel system of neuroendocrine peptidergic neurons in the pond snail, *Lymnaea stagnalis*, with antisera to the teleostean hormone hypocalcin and parathyroid hormone. *General and Comparative Endocrinology* 75 29-38.

84   Wikgren M (1990) The neuroendocrine system of flatworms. In *Progress in Comparative Endocrinology*, pp 323-328. Eds A Epple, CG Scanes & MT Stetson. New York: Wiley-Liss.

85   Yao CZ, Ishizuka J, Townsend CM & Thompson JC (1993) Successful intracerebroventricular allotransplantation of parathyroid tissue in rats without immunosuppression. *Transplantation* **55** 251-25.

Note added in proof

The Ca^{2+}-sensing receptor that regulates parathyroid function in the periphery, has recently been found in the brain (86). This may indicate a relationship between neural PTH secretion and extracellular Ca^{2+}. The recent discovery that PTH-specific (PTH2) receptors are more abundant in the brain than in 'traditional' PTH-target tissues (87) also suggests that the brain is an important site for PTH action.

86   Brown EM, Gamba G, Riccardi D, Lombardi M, Butters R, Kifor O, Sun A, Hediger MA, Lytton J & Hebert SC (1993) Characterization of an extracellular Ca^{2+}-sensing receptor from bovine parathyroid. *Nature* **366** 575-580.

87   Usdin TB, Gruber C & Bonner TI (1995) Identification and functional expression of a receptor selectively recognizing parathyroid hormone, the PTH2 receptor. *Journal of Biological Chemistry* **270** 15455-15458.

The Comparative Endocrinology of Calcium Regulation
Eds C Dacke, J Danks, I Caple & G Flik, pp 17-42
Journal of Endocrinology Ltd, Bristol (1996)

An exploration of salt and water balance in invertebrate and vertebrate animals

P C Withers

Department of Zoology, University of Western Australia, Nedlands,
Western Australia 6907, Australia

Introduction

There have been numerous reviews and books written on the composition and regulation of the extracellular fluids in various specific groups of animals (see below for references), but it is timely to review here the general patterns of extracellular fluid composition and ionic regulation in animals from a modern and synthetic perspective.

This chapter first presents an overview of the patterns of extracellular fluid composition for a wide range of marine, freshwater and terrestrial animals, then addresses specific questions arising from this general survey, including the composition of 'ancient' seawater and extracellular body fluids of primitive marine animals, the composition of the extracellular fluids of freshwater animals, the possible freshwater origins of vertebrates, the ureo-osmoconforming strategy of some fish and frogs, various potential roles of nitrogenous solutes as osmolytes, the contribution of some solutes to positive buoyancy, and osmolyte compatibility with protein structure and function.

This review is restricted to patterns of extracellular solute composition amongst animals, although intracellular patterns of solute composition would be generally predictable from the particular extracellular pattern. There must be, at equilibrium, osmotic balance across the plasmalemma for animal cells, and so patterns of osmoregulation discerned from the overall solute composition of the extracellular fluids are equally applicable to the patterns of intracellular osmoregulation, although the specific solutes involved are invariably different.

Extracellular and intracellular fluid composition

The solute composition of seawater is essentially ionic (12), and predominated by sodium (470 mM) and chloride (548 mM) ions, with lower concentrations of magnesium (54 mM), potassium (10 mM), calcium (10 mM) and sulphate (28 mM) ions.

The 'major' ions of extracellular body fluids of animals (based on concentration, ≥ 50 mM or higher) are generally Na^+ and Cl^-, with K^+ being the 'major' intracellular cation (Table 1). Generally, Ca^{2+} and Mg^{2+} are 'minor' ions, although they have a wide variety of extremely important roles and can in some circumstances be present in 'major' concentrations. In some animals, various organic solutes accumulate to 'major' concentrations, especially amino acids (particularly glycine, proline, taurine), some polyhydric alcohols (polyols) such as glucose, trehalose and glycerol, and various nitrogenous solutes including urea and trimethylamine oxide (TMAO). Proteins are often present in 'major' concentrations based on mass, but their high molecular weights mean they have minor osmotic effects (although their membrane impermeability imparts significant Donnan effects).

Table 1 Comparison of the ionic composition of intracellular and extracellular fluids for representative animals, a marine invertebrate (horseshoe crab *Limulus*), a freshwater bony fish (salmon *Salmo*) and a freshwater invertebrate (mollusc *Anodonta*). Values are in mM. (Modified from (117).)

	Horseshoe crab		Salmon		Freshwater mollusc	
	Intracellular	Extracellular	Intracellular	Extracellular	Intracellular	Extracellular
Na^+	28.8	445	20.5	140	5.3	15.6
K^+	129	11.8	264	4.7	21.3	0.5
Cl^-	43.2	514	3.2	128	2.4	11.7

There are almost invariably substantial differences between the extracellular and intracellular fluid solute concentrations, based either on active transport and relative membrane impermeability e.g. Na^+ and K^+, or passive distribution across the membrane based on the trans-membrane potential e.g. Cl^-, but some solutes are in diffusional equilibrium e.g. urea. There is a higher extracellular Na^+ and Cl^- concentration, and a higher intracellular K^+ concentration in all animals, although the actual concentrations vary widely amongst animals, depending primarily upon their habitat and extent of iono- and osmo-regulation (see Table 1 and below). For example, the extracellular Na^+ and Cl^- concentrations are 3 to 16 times the intracellular concentrations in the marine horseshoe crab (*Limulus*) and a freshwater mollusc (*Anodonta*), although the intracellular Na^+ and Cl^- concentrations are actually twice as high for the horseshoe crab than the extracellular Na^+ and Cl^- concentrations for the freshwater mollusc (Table 1)! In addition, many organic solutes are preferentially retained and have much higher concentrations within cells (e.g. glucose, amino acids, proteins).

Patterns of extracellular solutes
A cursory perusal of a comprehensive summary of data for solute composition of animals (Table 2) indicates that most marine invertebrate animals, and the hagfish, have a solute composition similar to seawater, whereas the cartilaginous fish (sharks, rays, chimaeras) and the coelacanth are hypo-ionic but iso-osmotic, with a high concentration of urea, and bony fish are hypo-ionic and hypo-osmotic. Freshwater animals have much lower solute compositions than seawater, with some freshwater

Table 2 Solute composition for seawater and the extracellular body fluids of a variety of seawater (SW), brackish water (BW), freshwater (FW) and terrestrial (TER) animals (all indicates that no distinction was made between aquatic or terrestrial species). Values are mean±standard error, in mM, with the number of observations in parentheses.

		Na^+	K^+	Cl^-	Ca^{2+}	Mg^{2+}	SO_4^{2-}	Urea
Seawater		470	10	548	10	54	28	-
Cnidarians	SW	457 (1)	10 (1)	554 (1)	10 (1)	51 (1)	15 (1)	-
	FW	31 (2)	0.4 (2)	6.5 (1)	-	-	-	-
Echinoderms	SW	458±10 (15)	10.6±0.5 (15)	524±11 (15)	10.3±0.3 (15)	50.2±2.2 (15)	28.9±0.8 (13)	-
	BW	216 (1)	5.4 (1)	255 (1)	5.6 (1)	24.2 (1)	26.5 (1)	-
Sipunculids	SW	503 (2)	11.5 (2)	557 (2)	11 (1)	39 (1)	26.5 (1)	-
Rotifers	FW	21 (1)	7 (1)	-	-	-	-	-
Annelids	SW	463 (3)	13.3 (3)	545 (3)	10 (2)	52 (2)	25.5 (2)	-
(oligochaetes)	FW	70±5.3 (7)	6.3±1.1 (7)	44±7 (7)	3.8±0.6 (7)	10.2 (2)	-	-
(hirudines)	FW	130 (2)	5 (2)	36 (2)	-	-	-	-
(oligochaetes)	TER	68.5±5.0 (8)	7.5±1.6 (8)	54.9±7.0 (8)	6.7±1.5 (7)	6.4±0.7 (5)	-	-
Molluscs	SW	468±9 (17)	13.6±0.9 (17)	543±8 (15)	11.1±0.3 (17)	53.8±0.7 (15)	18.3±3.1 (8)	-
	FW	41.1±5.8 (16)	1.7±0.3 (16)	28.9±4.1 (15)	5.7±0.7 (14)	2.2±0.8 (9)	-	-
	TER	61.0±9.0 (9)	3.7±0.9 (9)	57.1±8.7 (9)	8.3±1.5 (9)	6.0±2.4 (8)	-	-
Onycophorans	TER	93 (2)	3.8 (2)	80 (2)	4.3 (2)	0.8 (2)	4.5 (1)	-
Crustaceans	SW	447±9(56)	11.4±0.5 (56)	485±9 (53)	15.5±0.8 (49)	27.2±4.1 (45)	16.0±2.7 (8)	-
(except isopods)	FW	230±18 (27)	6.2±0.3 (26)	216±16 (27)	10.8±1.9 (16)	5.0±1.0 (13)	-	-
	TER	332±21 (11)	8.9±1.3 (11)	333±22 (11)	16.0±1.3 (10)	19.3±5.7 (10)	-	-
Isopods	SW	595 (2)	20 (2)	670 (2)	60.5 (2)	74 (2)	-	-

Table 2 (continued) Solute composition for seawater and the extracellular body fluids of a variety of seawater (SW), brackish water (BW), freshwater (FW) and terrestrial (TER) animals (all indicates that no distinction was made between aquatic or terrestrial species). Values are mean±standard error, in mM, with the number of observations in parentheses.

		Na^+	K^+	Cl^-	Ca^{2+}	Mg^{2+}	SO_4^{2-}	Urea
	TER	280±27 (4)	11.3 (3)	321±30 (4)	36 (2)	29 (2)	-	-
Insects	all	56.0±5.4 (116)	29.3±1.7 (115)	54.0±7.9 (33)	12.8±1.1 (86)	29.5±3.2 (73)	-	-
Spiders	TER	83 (2)	4.9 (2)	70 (2)	1.8 (2)	-	-	-
Scorpions	TER	243 (3)	5.3 (3)	265 (3)	5.3 (3)	-	-	-
Hagfish	SW	479±33 (5)	9.0±0.5 (5)	468±34 (5)	5.2±0.4 (5)	14.9±2.9 (5)	5.2 (2)	2.8 (1)
Lamprey	FW	117±12 (5)	5.6±0.7 (5)	96.6±8.1 (5)	2.3±0.2 (4)	1.9±0.2 (4)	1.3±0.6 (4)	-
Cartilaginous fish	SW	265±71 (15)	10.9±2.9 (15)	270±72 (15)	5.4±1.6 (12)	2.2±0.7 (12)	0.6 (2)	314±91 (13)
	BW	214±16 (11)	7.3±0.8 (11)	195±11 (15)	5.8±0.8 (11)	2.4±0.3 (11)	1.1±0.5 (5)	171±37 (12)
	FW	150 (1)	5.9 (1)	149 (1)	7.2 (1)	3.6 (1)	-	0.5 (1)
Coelacanth	SW	197 (1)	5.8 (1)	187 (1)	5.1 (1)	5.3 (1)	4.8 (1)	377 (1)
Bony fish	SW	190±7 (19)	5.5±0.9 (17)	160±6 (19)	3.2±0.3 (16)	3.3±0.6 (15)	1.8±0.7 (8)	13.9±5.7 (7)
	FW	140±3 (18)	3.7±0.5 (18)	122±3 (18)	2.9±0.4 (15)	1.3±0.2 (11)	1.4±0.4 (6)	4.5±2.3 (4)
Amphibians	FW	113±7 (8)	4.4±0.6 (8)	79±4 (7)	2.1±0.4 (4)	2.1±0.8 (4)	-	-
	all	142±12 (17)	5.1±0.7 (17)	117±16 (15)	2.2±0.3 (6)	1.7±0.5 (6)	2 (1)	175±90 (6)
Reptiles	all	151±3 (61)	4.9±0.2 (61)	11.8±3 (60)	5.6±2.4 (36)	2.2±0.5 (30)	0.6±0.1 (8)	15.7±10.8 (11)
Birds	all	157±1 (16)	3.0±0.2 (17)	118±2 (17)	3.4±1.3 (3)	2.0 (1)	-	7.8±1.0 (14)
Mammals	all	153±3 (17)	5.3±0.2 (18)	110±2 (16)	3.2±0.3 (13)	1.2±0.2 (12)	0.9 (2)	9.4 (2)

Data from (in part) references 2, 4, 6, 7, 8, 9, 10, 17, 19, 24, 26, 27, 35, 36, 37, 46, 50, 52, 55, 56, 58, 59, 60, 71, 74, 77, 80, 81, 82, 85, 88, 90, 93, 98, 100, 103, 106, 108, 109, 111, 114, 117, 119, 120.

molluscs having a remarkably dilute extracellular fluid. Terrestrial crustaceans have an extracellular solute composition more similar to seawater than most other terrestrial animals, which generally have solute compositions more like their freshwater counterparts. The group of animals which appears most unusual is the insects, which generally have low Na^+ and Cl^- concentrations and high K^+ and Mg^{2+} concentrations.

This considerable body of data can be graphically presented and quantitatively analysed for general patterns using principal component analysis (PCA). For a general summary, I first analysed by PCA (using SYSTAT; 116) only the major extracellular ions (Na^+, Cl^-), K^+ (the major intracellular ion) and urea (urea concentrations were assumed to be 0 mM for all animals except those for which concentrations in excess of 30 mM had been measured). This PCA analysis provided three factors (PCA axes 1, 2 and 3) which separated the various animals based on combinations of their extracellular concentrations of Na^+, K^+, Cl^- and urea; these three PCA factors each have biological significance.

PCA factor 1 is equal to $0.481[Cl^-]+0.476[Na^+]+0.177[K^+]+0.024[urea]$, where the solute concentrations are the difference from the standardised mean values for all of the data (Cl^-, 220±sd 178 mM; Na^+, 224±sd 150 mM; K^+, 8.3±sd 7.1 mM; urea, 20.1±sd 76.6 mM; (sd is the standard deviation) $n=409$). For example, the calculated value of PCA factor 1 for seawater is

$$0.481\left(\frac{548-220}{178}\right) + 0.476\left(\frac{470-224}{150}\right) + 0.177\left(\frac{10-8.3}{7.1}\right) + 0.024\left(\frac{0-20.1}{76.6}\right) = 1.72$$

PCA factor 1 explains more of the variance amongst the data (51%) than factors 2 and 3. The predominant solutes contributing to PCA factor 1 are the major extracellular ions, Na^+ and Cl^-. Therefore, the predominant pattern for various animals is a fairly linear change from seawater animals (high Na^+ and Cl^-) to freshwater animals (low Na^+ and Cl^-) along the PCA factor 1 axis, with brackish water animals being intermediate and freshwater cnidarians and molluscs at the extreme low end of the range (Fig. 1). This is most clearly seen in the panel of Fig. 1 for vertebrate amniote animals (reptiles, birds and mammals) and all other invertebrate animals, where most animals fit on an essentially straight, horizontal line, with almost no variability along the factor 2 axis; only some fish and amphibians, and insects, substantially deviate from this line, along PCA axis 2 (Figure 1, bottom panel).

PCA factor 2 is $0.040[Cl^-]+0.042[Na^+]-0.091[K^+]-0.994[urea]$, which explains a further 25% of the variance in the solute data. Clearly, the major solute contributing to factor 2 is urea, and so cartilaginous fish, the coelacanth and some terrestrial and seawater amphibians are separated from the other animals on this axis (lower panel, Fig. 1).

PCA factor 3 is $0.152[Cl^-]+0.213[Na^+]-0.998[K^+]+0.106[urea]$. PCA factor 3 explains 23% of the variance in the solute data. Clearly, the major solute contributing to factor 3 is K^+, and many of the insects are separated from the other animals on this axis (see below).

The following discussion first interprets each of these PCA axes (i.e. Na^+/Cl^-, urea, K^+) as an evolutionary strategy for the regulation/conformation of the

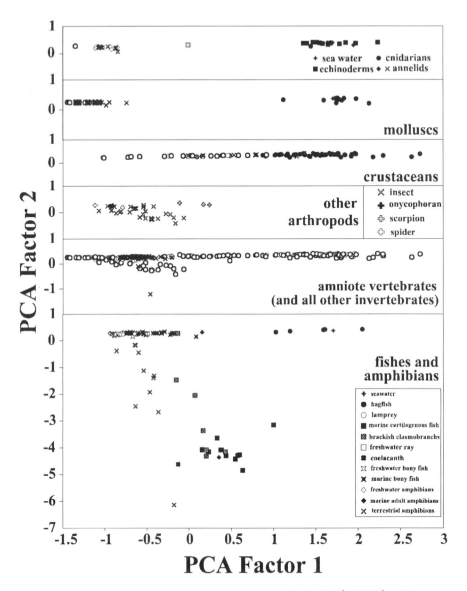

Fig. 1 Factors 1 and 2 for the principal component analysis of Na$^+$, Cl$^-$, K$^+$, and urea concentrations of animals. Each panel presents data for a different group of animals, and the panel for vertebrate amniote animals includes all invertebrate animals for comparison. For the top four panels, seawater animals are indicated by a solid symbol, brackish water animals by a grey symbol, freshwater animals by an open symbol, and terrestrial by a cross or plus. See text for further details.

extracellular body fluid solutes, and examines in further detail a number of considerations arising from these three main PCA axes: (1) marine animals; (2) seawater-brackishwater-freshwater transitions; (3) reinvasion of seawater from freshwater by some invertebrate and vertebrate animals; (4) terrestrial transitions, including insects with high K^+ and low Na^+ and Cl^-. Then, the general role of various nitrogenous solutes (including urea) as osmolytes, positive buoyancy solutes, and perturbing/compatible/counteracting solutes is discussed.

Marine animals

The solute composition of the extracellular fluids of marine animals might be expected to generally reflect the solute composition of the marine environment in which those animals evolved. When unicellular animals evolved about a billion years ago, 'ancient' seawater was their extracellular fluid. For primordial multicellular animals, the extracellular fluid space was essentially seawater (e.g. cnidarians, echinoderms). For more advanced multicellular animals, the extracellular fluid space became physically isolated from the external seawater environment, and there arose the ability to regulate the concentrations of some, or even all, solutes of the extracellular fluids. Nevertheless, many marine invertebrate animals (e.g. annelids, molluscs, crustaceans) retain an extracellular fluid composition very similar to seawater. Summary values for taxonomic groups have been provided in Table 2, and are shown for individual species by PCA in Fig. 1.

It is apparent that many marine invertebrate animals (as well as hagfish) have extracellular body fluids that generally resemble seawater whereas other marine vertebrates have lower ion concentrations. Nevertheless, there are significant differences in extracellular solutes amongst taxonomic groups of marine invertebrate animals (and hagfish) and vertebrates. Marine isopods generally have significantly higher ion concentrations than the other marine invertebrate animals and hagfish, which tend to be more similar. However, there are significant differences in Cl^- between molluscs, crustaceans and echinoderms. There are significant differences in Ca^{2+} between crustaceans (high concentrations) compared with echinoderms, molluscs and hagfish; and between hagfish (low concentrations) compared with echinoderms and molluscs. Crustaceans may have high Ca^{2+} concentrations because of the role of Ca^{2+} in calcification of the exoskeleton. There are significant differences in Mg^{2+} between crustaceans (low concentrations) compared with echinoderms, hagfish and echinoderms, and hagfish (low concentrations) compared with molluscs. The low Ca^{2+} and Mg^{2+} concentrations of hagfish may reflect a freshwater ancestry (see below).

If the extracellular solute composition of many invertebrate animals now closely resembles that of present day seawater, how constant has the composition of seawater, and these animals, been over geologic time? The general consensus is that a steady-state solute composition of seawater may have been attained about 1.5 billion years ago, and that various ion sources and sinks, and sedimentary cycling have since then maintained a relatively stable seawater composition (64, 72). Nevertheless, there may have been minor perturbations in global seawater composition. For example, about $5\text{-}6 \times 10^6$ years ago, the Mediterranean Sea experienced a number of isolations from the global seas and was transformed into a series of salt lakes. This not only eliminated the marine biota of the Mediterranean Sea, but also removed so much salt from the global seas that their salinity may have declined by about 5%. Despite the supposed constancy of seawater composition, Spaargaren (107) attempted to determine the composition of 'ancient' seawater from the similarities in composition of various animal physiological salines. 'Ancient' seawater was estimated to be considerably less concentrated than seawater, though with a similar proportion of ions, perhaps through choice of physiological salines inappropriate for primitive (marine) animals.

Although there are some minor differences, the ionic and osmotic composition of marine invertebrate animals and hagfish is generally similar to seawater, i.e. they essentially ionoconform and osmoconform. This minimises any energetic costs of maintaining ion concentration gradients between seawater and the extracellular body fluids (86, 117).

Seawater-brackish-freshwater transitions

Animals which moved from the marine environment to brackish environments can still ionoconform and osmo-conform like their marine counterparts, although at lower concentrations. This strategy avoids an energetic cost of iono- and osmo-regulation. For example, the echinoderm *Asterias rubens* iono- and osmoconforms in both seawater (North Sea, ambient $Na^+=429$ mM, body fluid $Na^+=428$ mM) and brackish water (Kiel Sea, ambient $Na^+=215$ mM, body fluid $Na^+=216$ mM; refs 16, 99). The anemone *Diadumene leucolena* osmoconforms over a very wide range of brackish water, from 20 to 100% seawater (82).

An iono/osmo-conformation strategy would result in a continual fluctuation of the extracellular solute composition in continuously-varying environments, such as estuaries. For example, the triclad turbellarian worm *Gunda ulvae* lives in estuaries between tide marks, and is exposed twice per day to a medium changing from freshwater to seawater (65). At greater than about 70% seawater, *Gunda* iono- and osmo-conforms, but at lower concentrations it iono- and osmo-regulates, and even in freshwater it keeps its salt concentrations at about 5-10% of those in seawater. However, there is considerable variation amongst brackish water animals in the solute concentrations of the water at which they switch from iono- and osmo-conformation to iono- and osmo-regulation, and in the actual ionic and osmotic concentrations that they regulate.

For freshwater animals, the body fluid environment must be (and is) invariably maintained by iono- and osmo-regulation above the ambient water concentrations; the body fluids are hyperionic and hyperosmotic to freshwater but considerably lower than values for seawater (e.g. annelids, molluscs, crustaceans; Table 2). The reduction in body fluid solute concentrations reduces the metabolic cost of ionoregulation of the extracellular fluid (86, 117). Freshwater sponges, cnidarians, rotifers and some molluscs have very low ionic and osmotic concentrations (Table 2).

Reinvasion of seawater

For fresh or brackish water animals which returned to seawater over evolutionary time, or return during part of their life cycle by migration, the possible patterns of extracellular fluid solute regulation include: the return to essentially seawater composition (i.e. the primitive condition); the return towards seawater composition but nevertheless hypo-ionic and hypo-osmotic regulation; the adoption of urea to fill the 'solute gap' between body ion concentrations and the total osmotic concentration of seawater; the maintenance of similar extracellular fluid ion concentrations as in freshwater. The former two strategies would seem to be characteristic of invertebrate animals whereas the latter two strategies are characteristic of vertebrate animals.

Invertebrate animals

Some crustaceans (shrimp, brine shrimp, isopods) exhibit hypo-ionic and hypo-osmotic regulation, when migrating between brackish and seawater, or in hyperosmotic media (e.g. salt lakes). For example, the euryhaline shrimp *Palaemonetes varians* is normally iso-osmotic with 65% seawater but is a good hypo-ionic regulator in seawater (87). The brineshrimp, *Artemia salina*, is an exceptional hypoionic regulator and survives in even exceedingly concentrated brine solutions (29). The isopod *Haloniscus searlei*, which belongs to an almost exclusively terrestrial taxon of isopods, is an extremely good osmoregulator over 3% to 460% seawater, being hypo-osmotic at >54% seawater (14). Other common hypersaline invertebrate animals include various larval and adult insects.

Vertebrate animals

There is considerable debate about whether vertebrates evolved in freshwater or seawater (47-49). Based on patterns of iono- and osmo-regulation and the structure and the functioning of vertebrate kidney nephrons, a freshwater origin is generally considered to be consistent for all vertebrates except for the anomalous hagfish, which are marine and essentially ionoconform and osmoconform, unlike any other vertebrates. If the first vertebrates were freshwater animals, then all marine vertebrates (hagfish, lampreys, cartilaginous fish, bony fish, coelacanths and a few semi-marine frogs, as well as many reptiles and mammals) have returned to seawater. Alternatively, the first vertebrates may have been marine, then the jawed vertebrates moved to freshwater, and some subsequently returned to seawater. In this case, the lamprey might have originally been marine, and evolved a freshwater breeding and larval development stage, or might have evolved from a jawless freshwater vertebrate stock and returned to seawater for the feeding adult life stage.

The hagfish are certainly exceptional marine vertebrates, in essentially iono- and osmo-conforming with seawater. However, a principal component analysis of the 'minor' ions (K^+, Ca^{2+}, Mg^{2+} and SO_4^{2-}) for marine animals (Fig. 2) clearly separates vertebrates (including hagfish) from invertebrate animals, despite hagfish basically being ionoconformers and osmoconformers like the marine invertebrate animals. PCA factor 1, which explained 74% of the variance, was $0.261[K^+]+0.297[Ca^{2+}]+0.319[Mg]+0.284[SO_4^{2-}]$, with mean$\pm$sd values as follows: K^+, 9.5±4.4; Ca^{2+}, 8.8±4.0; Mg^{2+}, 31.9±22.9; SO_4^{2-}, 16.0±11.8 (n=46). Thus, factor 1 was a fairly uniform mix of all ions used in the analysis. The second PCA factor, which explained 16% of the variance, was $-0.955[K^+]+0.045[Ca^{2+}]+0.040[Mg]+0.785[SO_4^{2-}]$; i.e. primarily K^+ and SO_4^{2-}. The third PCA factor explained less than 10% of the variance. A PCA using just K^+, Ca^{2+} and Mg^{2+} (not shown) also places hagfish amongst the other marine vertebrates. Again, this similarity of ionoregulation of the 'minor' ions for hagfish and other marine vertebrates, despite the great dissimilarity in their Na^+ and Cl^-, may reflect a freshwater ancestry of hagfish. Hagfish have several other 'freshwater vertebrate' features (48, 49), including glomerular kidneys (105), relatively salt-impermeable gills and skin (76), gill ion pumps for Na^+/Cl^- uptake in exchange for H^+/HCO_3^- (34) and typical vertebrate chloride cells (13, 73).

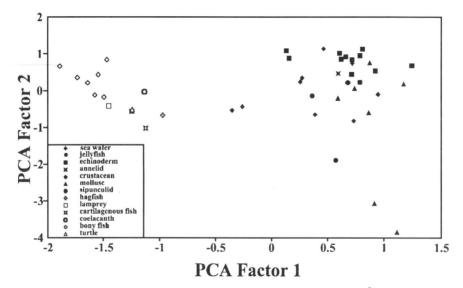

Fig. 2 Factors 1 and 2 for the principal component analysis of K^+, Ca^{2+}, Mg^{++} and SO_4^{2-} for aquatic animals, showing the distinction between vertebrate animals (including the osmoconforming hagfish) and marine invertebrate animals. See text for further details.

It would appear that many aspects of ionoregulation by hagfish are consistent with a freshwater origin, and only the strategy of osmoconforming, using Na^+ and Cl^-, is uncharacteristically 'un-vertebrate'. If indeed hagfish were originally freshwater vertebrates which have reinvaded seawater, why did they adopt this unusual iono/osmoregulatory strategy in comparison with other marine vertebrates which either ureo-osmoconform or hypo-iono/osmoregulate? Perhaps hagfish (and lampreys?) lack a complete and functional urea cycle and therefore could not ureo-osmoconform. A functional urea cycle clearly evolved in aquatic vertebrates, perhaps for acid-base regulation rather than to produce a nitrogenous waste product (5, 123). Although cartilaginous and many bony fish have a functional urea cycle, it is not clear whether hagfish and lampreys do (22, 78). Or, perhaps hagfish do not urea-osmoconform for the same reasons that marine bony fish do not (see below). Or, perhaps hagfish do not osmoregulate or regulate the major ions because of the high metabolic cost of this strategy; hagfish are, after all, sluggish, benthic animals that feed on polychaete worms or dead vertebrates.

Marine cartilaginous fish (sharks, rays, chimaeras), and the coelacanth *Latimeria chalumnae*, osmoconform with seawater but retain ion concentrations substantially lower than seawater, although the concentrations for Na^+, K^+, Cl^- and Ca^{2+} are significantly higher than for marine bony fish (Table 2). The osmotic sum of the blood ions ($[Na^+]+[K^+]+[Cl^-]+[Ca^{2+}]+[Mg]+[SO_4^{2-}]$) of cartilaginous fish is about 550 mOsm whereas the total osmotic concentration is about 1050 mOsm; there is consequently an 'osmotic gap' of about 500 mOsm. Much of this osmotic gap is filled by urea, which accumulates to concentrations of about 320 mM, and part is filled by TMAO which accumulates to concentrations of about 80 mM (58, 119, 120). Various other solutes fill the remainder of the osmotic gap. This iono/osmoregulatory strategy is termed ureo-osmoconformation. In contrast, the exclusively freshwater stingrays (*Potamotrygon*) have ion concentrations like bony fish and virtually no urea (0.5 mM (111)), and other 'fresh' and brackish water cartilaginous fish have concentrations intermediate between *Potamotrygon* and marine species. The coelacanth has slightly higher plasma ion concentrations than seawater bony fish, and high concentrations of urea and TMAO (51). The ureo-osmoconforming strategy has independently evolved in cartilaginous fish and the coelacanth (and also a few amphibians - see below).

Why do marine cartilaginous fish and the coelacanth ureo-osmoconform rather than iono/osmoregulate? Factors perhaps predisposing marine cartilaginous fish to the ureo-osmconforming strategy include the constancy of the marine environment, and their relatively large size, sluggish metabolism and reproduction via viviparity, ovoviviparity or a 'cleidoic' egg (47). Factors perhaps predisposing marine bony fish to an iono/osmoregulatory strategy include their tendency for euryhalinity and moving between freshwater and seawater, small size, active metabolism and reproduction by ovipary with a 'non-cleidoic' egg (47). The ureo-osmoconforming strategy clearly requires the capacity to synthesise, accumulate and tolerate high levels of urea, but this apparently does not preclude bony fish from the ureo-osmoconforming strategy. The urea cycle is at least latent, if not expressed, in teleost fish (some teleost larvae synthesise urea for a short developmental period and some teleosts have a functional urea cycle (78, 79, 95, 124)). Other vertebrates, such as terrestrial or aestivating amphibians and lungfish, can also tolerate high urea concentrations (see below).

There are a few hypersaline fish, which invariably iono/osmoregulate at typical concentrations for marine fish. For example, the estuarine atherinid, *Leptatherina wallacei*, survives in ion concentrations more than twice those of seawater but maintains plasma Na^+ concentration less than 300 mM (110). An unusual cichlid fish, *Oreochromis alcalicus*, survives in alkaline hypersaline lakes; it excretes high concentrations of urea (89), presumably to maintain acid-base balance in its highly alkaline environment (5). A variety of other freshwater or seawater fish also synthesise and excrete significant levels of urea (78, 79, 95).

Amphibians are generally restricted to freshwater environments (or terrestrial environments with access to freshwater). However, a few tolerate high salinity. The crab-eating frog *Rana cancrivora* is found in brackish water approaching the concentrations of seawater, and adults use urea as a major balancing solute, although tadpoles iono- and osmo-regulate (44, 45). The African clawed frog (*Xenopus laevis*) also accumulates substantial levels of urea when acclimated to salinities up to 33% seawater (63, 94). The Brazilian intertidal frog, *Thoropa miliaris*, osmoconforms but does not appear to accumulate urea (1). Thus, accumulation of high urea concentrations has independently arisen in a few disparate amphibians, as a ureo-osmoconforming strategy. A number of dehydrated/aestivating frogs and salamanders also accumulate substantial concentrations of urea (see below), as does the aestivating lungfish *Protopterus æthiopicus*.

Terrestrial transitions

The transition from an aquatic to a terrestrial existence has involved many changes to the avenues and mechanisms for ionoregulation and osmoregulation (88, 117). Essentially, the concentrations of all solutes must be regulated through intake or excretion by terrestrial animals, as ions cannot be in passive equilibrium with the surrounding medium (air).

It might be expected that the solute concentrations regulated by terrestrial animals would reflect the solute concentrations of their aquatic ancestors, whether marine (high Na^+ and Cl^-) or freshwater (low Na^+ and Cl^-). However, it proves difficult to discern such evolutionary 'inertia' in patterns of solute regulation by terrestrial animals. The pattern of extracellular solute composition of some terrestrial animals is sometimes similar to seawater and the extracellular fluids of marine animals (e.g. many terrestrial crustaceans), but for most animals it is similar to the extracellular fluids of freshwater counterparts (e.g. terrestrial molluscs, most vertebrates).

Annelids

The extracellular solute composition of the semi-terrestrial oligochaete worms generally resembles that of freshwater species, although the ionic and osmotic concentrations are somewhat higher for worms maintained in soil with limited moisture than for worms maintained in freshwater (80). The blood Na^+ and Cl^- concentrations are generally about 69 and 57 mM respectively (range about 40 to 80 mM for both Na^+ and Cl^-), which are very similar to the values for freshwater annelids (Fig. 1, Table 2).

Molluscs

It might be expected that the seawater or freshwater ancestry of terrestrial snails would be reflected in their pattern of haemolymph solutes, since marine and freshwater snails differ more markedly in their extracellular solutes than other major animal taxa found in both environments. However, the extracellular solutes of terrestrial prosobranch snails, which have evolved from freshwater prosobranchs (96) are quite variable (Na^+ from 27 to 110 mM, Cl^- from 24 to 106 mM (24)), and similar to the extracellular solutes of terrestrial pulmonate snails (Na^+ from 35 to 68 mM, Cl^- from 53 to 72 mM), which may have evolved from estuarine or marine pulmonate snails. Thus, both prosobranch and pulmonate snails are similarly variable in extracellular solute concentrations, and both are markedly hypo-ionic and hypo-osmotic to seawater.

Onycophorans

The onycophorans, *Peripatus*, are clearly derived from marine ancestors (96) yet they have quite dilute haemolymph, with low osmotic and solute concentrations (Na^+ 93, K^+ 3.8, Cl^- 80, 200 mOsm; Table 2 (26, 93)), unlike terrestrial isopods and crabs. Perhaps terrestrial onycophorans are derived from an intermediary brackish or freshwater ancestor, rather than directly from seawater.

Crustaceans

A number of crab lineages are amphibious or have limited terrestriality, but their extracellular solute composition tends to be more like marine crustaceans than freshwater crustaceans (see Fig. 1; 46). Many 'terrestrial' crabs are essentially coastal species that are still dependent on seawater, and their extracellular Na^+ is about 400-460 mM and Cl^- is about 400-500 mM, similar to but often somewhat lower than, their seawater crab counterparts. Some intertidal and splash-zone species osmoconform (e.g. pagurids) or have a limited ability to iono/osmoregulate (e.g. the diogenid *Clibanarius*) but others (*Uca, Goniopsis*) iono/osmoregulate strongly with haemolymph concentrations lower than seawater. Species of *Cardiosoma*, which require contact with water in their burrows, are strong hypo-iono/osmoregulators in seawater; some species require freshwater whereas others can tolerate seawater. The more terrestrial hermit crabs regulate their haemolymph and shell water composition at 700-1300 mOsm, depending upon the species, e.g. *Coenobita brevimanus* 730-835 mOsm (haemolymph) and 835 mOsm (shell water) compared with *C. perlatus* 1107-1336 mOsm (haemolymph) and 1013-1316 mOsm (shell water). The terrestrial *Gecarcinus* (Na^+=459 mM) and *Gecarcoides* (Na^+=315-421 mM), *Geograpsus* (Na^+=367 mM) and *Ocypoda* (except *O. saratan*) osmoregulate slightly lower than seawater, whereas the freshwater/terrestrial crabs (e.g. *Sudanonautes, Potamon* and *Holthuisana*) have a considerably lower Na^+ (about 275 mM) and Cl^- (about 250 mM). The robber crab *Birgus latro* hypo-regulates the haemolymph Na^+ at 324 mM.

The most terrestrial group of crustaceans, the isopods, includes species that are strictly marine, strictly freshwater and terrestrial (114). The terrestrial isopods clearly have considerably lower ionic and osmotic concentrations than the marine isopods (Table 2; the differences for Na^+ and Cl^- are significant at $P<0.05$). For example, *Oniscus asellus* has a haemolymph composition like that of the terrestrial freshwater crustaceans mentioned above, of 255 mM Na^+, 20 mM K^+ and 256 mM Cl^- (108). Nevertheless, there is considerable variation amongst the terrestrial species.

Spiders and scorpions
Spiders have considerably lower haemolymph Na^+ and Cl^- concentrations (70-85 mM) than most terrestrial animals, whereas scorpions have considerably higher ion concentrations than spiders (Table 2).

Insects
The mean solute concentrations for insects (Table 2; values are mostly for terrestrial species but include some values for freshwater species) are quite unusual for animals, with very low Na^+ (56 mM) and Cl^- (54 mM), and unusually high K^+ (29 mM) and Mg^{2+} (29.5 mM). Such extracellular (haemolymph) ion concentrations are not typical of all insects, but reflect unusual values for particularly herbivorous larvae. For example, Sutcliffe (108) characterises insects into four categories.

(i)   Type I insects have mainly Na^+ and Cl^- as extracellular ions, with little K^+, Mg^{2+}, Ca^{2+} and PO_4^{2-} and free amino acids; Odonata, Ephemeroptera, Plecoptera, Dictyoptera and Hemiptera-Heteroptera are characteristically of this type.

(ii)   Type II insects are similar to type 1, except that Na^+ and Cl^- contribute less (each about 25%) to the total solute concentration, with high concentrations of largely unidentified solutes making up a substantial part of the balance; some Orthoptera, Dermaptera and Isoptera are of this type.

(iii)   Type III insects have a very low concentration of Cl^- relative to Na^+ (which can be 20-50% of the total solutes), with small amounts of K^+, Mg^{2+}, Ca^{2+} and PO_4^{2-} and a relatively high concentration of free amino acids (up to 25%); larvae of Trichoptera and Diptera, and Coleoptera, are of this type.

(iv)   Type IV insects have low Na^+ and Cl^- concentrations (<10% for each) with low amounts of K^+, Ca^{2+} and PO_4^{2-} and often a high Mg^{2+} and very high amino acid concentration; Lepidoptera and Hymenoptera are characteristically of this type. However, Sutcliffe (108) indicates that there is not a strict taxonomic pattern to these types.

It is clear from Table 2 and Fig. 1 that insects can differ markedly in $Na^+/Cl^-/K^+/$ urea composition from other terrestrial vertebrates, and that the difference is primarily in the K^+ concentrations (PCA factor 3). A principle component analysis of the ion composition for insects, using Na^+, K^+, Ca^{2+} and Mg^{2+} concentrations, indicates that 54% of the variation is explained by a PCA factor 1 of $0.367[K^+]-0.365[Na^+]+0.234[Ca^{2+}]+0.322[Mg^{2+}]$, i.e. a fairly equal contribution of all ions. A further 21% of the variance is explained by a PCA factor 2 of $0.423[Na^+]-0.326[K^+]+0.925[Ca^{2+}]+0.178[Mg^{2+}]$, i.e. mainly Ca^{2+}, and a further 14% by a PCA factor 3 of $0.475[K^+]-0.210[Na^+]+0.481[Ca^{2+}]-1.13[Mg^{2+}]$, i.e. mainly Mg^{2+}. The means and standard deviations for the solutes are; Na^+ 46.6±55.4, K^+ 30.5±19.1, Ca^{2+} 13.2±10.1, and Mg^{2+} 23.6±15.5 (n=57). Graphical examination of this PCA (Fig. 3) suggests two major groups of insects, one comprising types I, II and III, and the other group mainly comprising type IV, and two species (top right) with extremely high Ca^{2+} concentrations. As pointed out by Sutcliffe (108), the solute pattern 'types' are not strictly taxonomic. The types I, II and III insects generally fall within the cluster of points for all other animals (which were not included in this PCA analysis).

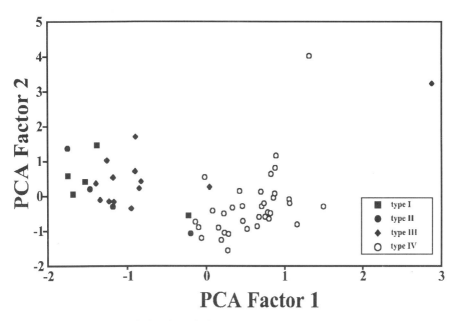

Fig. 3 Factors 1 and 2 for the principal component analysis of the major cations (Na^+, K^+, Ca^{2+} and Mg^{++}) in insects.

The unusual haemolymph composition of especially the type IV insects presumably reflects their phytophagous diet (which is relatively low in Na^+ and high in K^+, Ca^{2+} and Mg^{2+}), but does not imbalance the electrical potential of nerve cells because the nervous system is surrounded by a sheath of fat-body cells and the fluid in the extraneural space has a higher Na^+ concentration than the haemolymph (112, 117).

Vertebrate animals
Vertebrate animals (with the single exception of hagfish) regulate their extracellular ions at a fairly constant level, considerably less than seawater (although ureo-osmoconforming fish have slightly higher ion concentrations, and lampreys lower). There appears to be no substantial change in the pattern of ionoregulation or osmoregulation by vertebrates with the transition from aquatic to terrestrial habits, or the subsequent return of some to freshwater or seawater.

The ion concentrations of the primarily terrestrial amniote vertebrate animals (reptiles, birds, mammals) are remarkably similar, and intermediate between those of seawater and freshwater bony fish (Fig. 1, Table 2). Nevertheless, there are slight but significant differences in extracellular Na^+ and Cl^- (and sometimes urea) for many marine reptiles (turtles, crocodiles, snakes) compared with freshwater counterparts, and euryhaline species generally have higher ion concentrations in brackish or seawater (77). However, the blood ion concentrations of amphibians are lower than for other aquatic and terrestrial vertebrates (except lampreys, which also have low concentrations), and are presumably lower than those of their ancestors, regardless of whether those ancestors were marine, freshwater or brackish water inhabitants.

Positive buoyancy solutes

Mechanisms for adjustment of buoyancy by aquatic animals include an air-space (such as the swimbladder of teleost fish, cuttlebone of cephalopods, gas float of jellyfish), accumulation of low-density lipids (e.g. squalene in shark livers, fat droplets in many invertebrates), reduced calcification (e.g. cartilaginous fish) and use of hydrodynamic force to generate lift (117). For marine animals, the accumulation of specific solutes can also contribute to positive buoyancy.

It has long been appreciated that replacement of high molecular mass solutes by solutes of lower molecular mass reduces the density of body fluids, hence contributes to positive buoyancy. For example, body fluid density is reduced by the replacement of high molecular mass solutes, such as Ca^{2+} and SO_4^{2-} with lighter ions, such as NH_4^+, Na^+ and Cl^-. Replacement of SO_4^{2-} ions by lighter anions is especially apparent for cnidarians, but also for molluscs, crustaceans and vertebrates (Table 2). Ammonium can replace higher molecular mass solutes such as Na^+, and is used to lower the density of body fluids by some protozoans (*Noctiluca* (43), *Pyrocystis* (61)), tunicate eggs (66), squid (28, 30) and diatoms (53).

In recent years, the role of partial molal volume as a second strategy for density reduction has been elucidated (97, 119, 120). By changing the structure of water in a solution, and hence the solution density, certain solutes such as trimethylamine (TMA), TMAO and even urea can reduce solution density (Fig. 4). However, the use of special solutes for positive buoyancy control is relatively ineffectual, especially for freshwater animals and particularly in comparison with the use of air-filled spaces, because of the relatively small changes in density that can be achieved through the manipulation of solute composition. Therefore, either very high solute concentrations or very large solution volumes are required to achieve substantial changes in overall density by ion substitution. In freshwater, TMA is virtually the only solute that can confer any positive buoyancy; in seawater, TMA, TMAO, NH_4^+, Cl^-, urea, betaine and sarcosine can all contribute positive buoyancy (Fig. 4). TMA provides significant positive buoyancy to the marine crustacean *Notostomus* (97). In elasmobranchs, urea and TMAO can contribute significantly towards attainment of positive buoyancy due to their favourable partial molal volumes, although the overall contributions of these solutes do not allow the achievement of even positive buoyancy (119, 120).

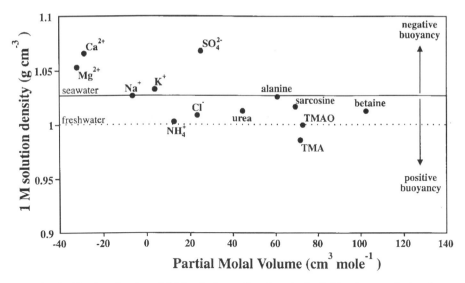

Fig. 4 Effective density of 1 M solutions of various body fluid solutes, calculated from their molecular mass and partial molal volume (119, 120).

Osmolytes

Most solutes present in body fluids have some physiological role and their concentration is determined by their specific function. However, in some circumstances particular solutes accumulate to high concentrations because of their osmotic potential, i.e. they act as osmolytes (57, 127). For example, urea is present at high concentrations (300-400 mM) in the body fluids of marine cartilaginous fish, the coelacanth, and certain amphibians (see above) and fills a large fraction of the 'solute gap' between the sum of the electrolyte concentrations and the total osmotic concentration required for osmoconformation with seawater. In some animals, urea also accumulates to high concentrations (often > 200 mM) during dehydration, but its role is not so much as an osmolyte *per se* but as a relatively non-toxic nitrogenous waste product that accumulates until normal excretion can be resumed. A number of dehydrated/aestivating frogs and salamanders also accumulate substantial concentrations of urea (33, 62, 68, 75, 101), as does the aestivating lungfish *Protopterus æthiopicus* (104). Other nitrogenous solutes that can function as osmolytes in a similar fashion as urea include TMAO, proline betaine, glycine betaine, glycero-phosphorylcholine (GPC) and free amino-acids. Polyhydric alcohols ('polyols') such as sorbitol, myoinositol, glycerol, glucose, trehalose, ethylene glycol, and erythritol, can also function as osmolytes.

The trimethylamine TMAO also accumulates in the extracellular fluids of marine elasmobranchs (85 mM (58, 119, 120)) and the coelacanth (122 mM (51)), filling a significant portion of the 'solute gap'. TMAO is also present at lower (≤1/10) concentrations in marine and a few freshwater bony fish (3). Embryos of the brine shrimp *Artemia* accumulate significant levels of TMAO/betaine during desiccation, but it is not clear whether it is TMAO, betaine, or a mixture of both (42). Whether TMAO accumulates as an osmolyte in other animals, such as amphibians at high salinities, or animal tissues, such as the mammalian and avian kidney, is apparently not known. However, TMAO does not accumulate to significant concentrations in aestivating Australian frogs (121). The high concentration of TMAO in elasmobranchs and the coelacanth appears to be important as a balancing osmolyte and perhaps as a counteracting solute (see below). TMAO can also protect proteins against thermal denaturation and is a cryoprotectant for frozen storage of tissues (3), but whether TMAO functions in either of these roles in animals is not known.

The role of glycine betaine (commonly called betaine) as an osmolyte was reported over 30 years ago (20, 21) but its role has been generally overlooked because of analytical difficulties. However, it has been shown to be regulated in response to hyperosmotic stress in oysters (84) and horseshoe crabs (31, 115). It accumulates to significant concentrations (>45 mM) in the renal medulla of certain mammals and birds (8, 15, 39, 67, 125). Glycine betaine is the major organic solute that accumulates during hypersaline exposure in the halotolerant bacterium *Spirulina* (38). The similar nitrogenous compound proline betaine has been identified as an important osmolyte in the mollusc *Elysia* (83). The polyamines spermine and putrescine may be involved in ionic and osmotic homeostasis of crabs by having direct affects on the Na-K-ATPase enzyme in gills (69). In contrast, betaine and GPC are only minor solutes in tissues of the clawed frog acclimated to about 1/3 seawater (122).

Amino-acids often accumulate in response to osmotic stress, particularly as intracellular osmolytes but also as extracellular osmolytes. For example, there are considerable increases in proline during hyperosmotic stress in the crab *Callinectes* (25, 41) and glycine, taurine and arginine in the crab *Carcinus* (128). In addition, amino-acids are important haemolymph osmolytes in many insects, particularly aquatic larvae in saline environments (37). For example, osmoconforming brackish water mosquito larvae accumulate various amino acids, especially proline and serine, and also the polyol, trehalose (40).

Polyols have recently been shown to accumulate in some animals. The mammalian renal medulla has high concentrations of inositol (over 20 mM) and sorbitol (50 mM), as well as betaine and GPC (8, 39, 125). In the avian kidney medulla, myoinositol accumulates during dehydration (along with betaine and taurine), but no sorbitol is present and GPC concentration does not alter (67). Trehalose is also an accumulated osmolyte in microorganisms (23) and brackish water mosquito larvae (40). However, polyols also have a major role in freeze avoidance and freeze protection (see (117) for a summary) because of the freezing point depression of high osmotic concentrations (e.g. over 2.5 M glycerol in the wasp *Bracon* and ethylene glycol in the beetle *Ips*).

Perturbing, compatible and counteracting solutes

Some organic solutes accumulate as osmolytes in extracellular/intracellular body fluids to substantial concentrations, and some ions (Na^+, Cl^-, K^+) are normally present in high concentrations. Solutes, particularly at high concentrations, can have physico-chemical effects on other solutes and water; these effects can be beneficial or detrimental. For example, some solutes in high concentrations bind non-specifically with proteins, and unfold and destabilise their structure (57, 70); these are perturbing solutes. Urea, arginine, Na^+, K^+ and Cl^- are perturbing solutes. In contrast, other solutes interact with proteins and stabilise their structure; these are compatible solutes. Some amino acids and glycerol are compatible solutes. Some solutes appear to specifically ameliorate the perturbing effects of other solutes and are called counteracting solutes. Cations such as NH_4^+, anions such as F^-, PO_4^{3-} and SO_4^{2-}, methylated amines (TMA, TMAO, betaine) and polyols (glucose, trehalose, sorbitol, etc.) are counteracting solutes.

Not surprisingly, compatible solutes rather than perturbing solutes are often used as balancing osmolytes by osmoconforming animals or for protection against intracellular water loss during dehydration, or for protection against freezing. For example, many marine invertebrate animals, hagfish and elasmobranchs accumulate amino acids as intracellular osmolytes to balance the osmotic pressure with seawater (92, 102, 123) and some insect larvae similarly accumulate amino acids (40).

Urea, in contrast, is a perturbing solute but it is often accumulated by animals as a balancing osmolyte (e.g. cartilaginous fish, coelacanths, some frogs) or during dehydration (e.g. aestivating lungfish, some amphibians and some molluscs). Urea is usually a nitrogenous end product in mainly vertebrate animals, but it and the other common nitrogenous end products (NH_4^+, uric acid) often assume different physiological roles (123). Regardless of its presumed 'perturbing' effects, urea is accumulated to high concentrations (>200 mM) by a number of vertebrates, sometimes without the concommitant accumulation of any counteracting solutes (see below), and so it seems to be less perturbing than is generally assumed. Studies of the physiological effects of urea in urea-accumulating animals and other animals would seem a fruitful avenue for future study.

It has been suggested that some compatible solutes also specifically ameliorate the inimical effects of other perturbing solutes (e.g. urea) and are therefore also counteracting solutes. For example, TMAO (and also betaine, sarcosine, alanine and taurine to a progressively lesser extent) has the opposite effect of urea on the catalytic properties of some enzymes of some cartilaginous fish (57, 126), and actually maintains catalytic constancy when present in a 1:2 ratio with urea. The ratio of TMAO:urea is often about 1:2 for extracellular fluids of elasmobranchs, and the intracellular ratio is generally about 1:2 for total methylamines:urea. However, a counteracting effect of TMAO is not observed for all elasmobranchs, so it is unclear whether TMAO generally has a counteracting effect in addition to its role as a compatible solute. In the marine cyanobacterium *Spirulina*, betaine protects enzyme activity (K_m and V_{max}) at high ionic concentrations, and glycerol and proline counteract effects on V_{max} (but not K_m (38)).

Aestivating lungfish (*Protopterus*) and various amphibians accumulate substantial levels of urea in body fluids during dehydration, by accelerated rates of urea synthesis and/or reduced renal excretion (105, 101). The African clawed frog, *Xenopus laevis*, accumulates substantial levels of urea (up to 100-150 mM) as an osmolyte when acclimated to high concentrations of NaCl (63, 94). This ureo-osmoconformation presumably mimics urea accumulation during aestivation (11). However, ureo-accumulation up to 80 mM is not accompanied by significant accumulation of the counteracting solutes betaine or GPC (122). Ureo-accumulation by aestivating Australian desert frogs (*Neobatrachus, Cyclorana*), even at over 200 mM levels, is not accompanied by significant accumulation of TMAO, betaine, sarcosine or GPC (121). It is not known whether other aestivating amphibians or lungfish, or the marine crab-eating frog *Rana cancrivora,* accumulate TMAO or other counteracting solutes. There does not appear to be a perturbing effect of urea or a counteracting effect of TMAO in frogs. In fact, urea appears to be a non-perturbing osmolyte that might ameliorate the effects of increased ionic strength on enzyme function (54). Perhaps it is not so important to use counteracting solutes to maintain enzyme function during the state of metabolic depression that accompanies aestivation (104, 118) compared with the steady-state ureo-osmoconforming strategy of cartilaginous fish and the coelacanth. Or, perhaps urea is not so perturbing as is generally assumed and it is not universally necessary to co-accumulate 'counteracting' solutes.

References

1   Abe AS & Bicudo JEPW (1991) Adaptations to salinity and osmoregulation in the frog *Thoropa miliaris* (Amphibia, Leptodactylidae). *Zoologischer Anzeiger* **227** 313-318.

2   Altman PL & Dittmer DS (1961) *Blood and Other Body Fluids.* Washington, DC: Federation of American Societies for Experimental Biology.

3   Anthonini U, Børresen T, Christophersen C, Gram L & Nielsen PH (1991) Is trimethylamine oxide a reliable indicator for the marine origin of fish? *Comparative Biochemistry and Physiology* **97B** 569-571.

4   Arad Z, Horowitz M, Eylath U & Marder J (1989) Osmoregulation and body fluid compartmentalization in dehydrated heat-exposed pigeons. *American Journal of Physiology* **257** R377-382.

5   Atkinson DE (1992) Functional roles of urea synthesis in vertebrates. *Physiological Zoology* **65** 243-267.

6   Bakker HR & Bradshaw SD (1983) Renal function in the spectacled hare-wallaby, *Lagorchestes conspicillatus*: effects of dehydration and protein deficiency. *Australian Journal of Zoology* **31** 101-108.

7   Bakker HR, Bradshaw SD & Main AR (1982) Water and electrolyte metabolism of the Tammar wallaby *Macropus eugenii. Physiological Zoölogy* **55** 209-219.

8   Balaban RS & Burg MB (1987) Osmotically active organic solutes in the renal inner medulla. *Kidney International* **31** 562-564.

9   Balasch J, Palacios L, Musquera S, Palomeque J, Jiménez M & Alemany M (1973) Comparative hematological values of several galliformes. *Poultry Science* **52** 1531-1534.

10   Balasch J, Palomeque J, Palacios L, Musquera S & Jiménez M (1974) Hematological values of some great flying and aquatic-diving birds. *Comparative Biochemistry and Physiology* **49A** 137-145.

11   Balinsky JB, Choritz EL, Coe CGL & Van der Schans GS (1967) Amino acid metabolism and urea synthesis in naturally aestivating *Xenopus laevis*. *Comparative Biochemistry and Physiology* **22** 59-68.

12   Barnes HJ (1954) Ionic composition of sea water. *Journal of Experimental Biology* **31** 582-588.

13   Bartels H (1985) Assemblies of linear arrays of particles in the apical plasma membrane of mitochondria-rich cells in the gill epithelium of the Atlantic hagfishes (*Myxine glutinosa*). *Anatomical Record* **211** 229-238.

14   Bayly IAE & Ellis P (1969) *Haloniscus searlei* Chilton: an aquatic 'terrestrial' isopod with remarkable powers of osmotic regulation. *Comparative Biochemistry and Physiology* **31** 523-528.

15   Bedford JJ, Smiley M & Leader JP (1995) Organic osmolytes in the kidney of domesticated red deer, *Cervus elephas*. *Comparative Biochemistry and Physiology* **110A** 329-333.

16   Binyon J (1962) Ionic regulation and mode of adjustment to reduced salinity of the starfish *Asterias rubens* L. *Journal of the Marine Biological Association of the United Kingdom* **42** 49-64.

17   Binyon J (1966) Salinity tolerance and ionic regulation. *Physiology of Echinodermata* pp 359-377. Ed R A Boolootian. New York: Interscience Publishers.

18   Binyon J (1972) *Physiology of Echinoderms*. Oxford: Pergamon Press.

19   Bradshaw SD & Rice GE (1981) The effects of pituitary and adrenal hormones on renal and postrenal absorption of water and electrolytes in the lizard, *Varanus gouldii* (Gray). *General and Comparative Endocrinology* **44** 82-93.

20   Bricteux-Gregoire S, Ducateau-Bosson Gh, Jeuniaux Ch & Florkin M (1964) Constituants osmotiquement actifs des muscles adducteurs d'*Ostrea edulis* adaptée à l'eau de mer ou à l'eau saumâtre. *Archives Internationales de Physiologie et de Biochimie* **72** 267-275.

21   Bricteux-Gregoire S, Ducateau-Bosson Gh, Jeuniaux Ch & Florkin M (1964) Constituants osmotiquement actifs des muscles de *Gryphaea angulata* adaptée à l'eau de mer ou à l'eau saumâtre. *Archives Internationales de Physiologie et de Biochimie* **72** 835-842.

22   Brown GW & Brown SG (1985) On urea formation in primitive fishes. *Evolutionary Biology of Primitive Fishes*, pp 321-337. Eds RE Foreman, A Gorbman, JM Dodd & R Olsson. New York: Plenum Press.

23   Brown AD, Mackenzie KF & Singh KK (1986) Selected aspects of microbial osmoregulation. *EMBO-FEMS Workshop on the Molecular Basis of Haloadaptation in Microorganisms*, pp 32-36. Eds WD Grant & M Kogut. Amsterdam: FEMS Microbiology Series.

24   Burton RF (1983) Ionic regulation and water balance. *The Mollusca, Volume 5 Physiology, Part 2*, pp 291-352. Eds ASM Saleuddin & KM Wilbur. New York: Academic Press.

25   Burton RS (1992) Proline synthesis during osmotic stress in megalopa stage larvae of the blue crab, *Callinectes sapidus*. *Biological Bulletin* **182** 409-415.

26   Campiglia SS (1976) The blood of *Peripatus acacioi* Marcus & Marcus (Onycophora)-III. The ionic composition of the haemolymph. *Comparative Biochemistry and Physiology* **54A** 129-133.

27   Clark ME (1968) Free amino-acid levels in the coelomic fluid and body wall of polychaetes. *Biological Bulletin* **134** 35-47.

28   Clarke MR, Denton EJ & Gilpin-Brown JB (1979) On the use of ammonium for buoyancy in squids. *Journal of the Marine Biological Association of the United Kingdom* **59** 259-276.

29   Croghan PC (1958) The osmotic and ionic regulation of *Artemia salina* (L.). *Journal of Experimental Biology* **35** 219-233.

30   Denton EJ, Gilpin-Brown JB & Shaw TI (1969) A buoyancy mechanism found in cranchid squid. *Proceedings of the Royal Society of London B* **174** 271-279.

31   Dragovich J & Pierce SK (1992) Comparative time courses of inorganic and organic osmolyte accumulation as horseshoe crabs (*Limulus polyphemus*) adapted to high salinity. *Comparative Biochemistry and Physiology* **102A** 79-84.

32   Englehardt FR & Dehnel PA (1973) Ionic regulation in the Pacific edible crab *Cancer magister* (Dana). *Canadian Journal of Zoology* **51** 735-743.

33   Etheridge K (1990) Water balance in estivating sirenid salamanders (*Siren lacertina*). *Herpetologica* **46** 400-406.

34   Evans DH (1984) Gill Na^+/H^+ and Cl^-/HCO_3^- exchange systems evolved before the vertebrates entered freshwater. *Journal of Experimental Biology* **113** 465-469.

35   Fleming WR & Hazelwood DH (1967) Ionic and osmotic regulation in the fresh water medusa, *Craspedacusta sowerbyi. Comparative Biochemistry and Physiology* **23** 911-915.

36   Florkin M (1960) Blood chemistry. *The Physiology of Crustacea* pp 141-159. Ed T H Waterman. New York: Academic Press.

37   Florkin M & Jeuniaux C (1964) Hemolymph: composition. *The Physiology of Insecta* pp 110-152. Ed M Rockstein. New York: Academic Press.

38   Gabbay-Azaria R, Tel-Or E & Schönfeld M (1988) Glycine betaine as an osmoregulant and compatible solute in the marine cyanobacterium *Spirulina subsalsa. Archives of Biochemistry and Biophysics* **264** 333-339.

39   Garcia-Perez A & Burg MB (1991) Renal medullary organic osmolytes. *Physiological Reviews* **71** 1081-1115.

40   Garrett MA & Bradley TJ (1989) Extracellular accumulation of proline, serine and trehalose in the haemolymph of osmoconforming brackish-water mosquitos. *Journal of Experimental Biology* **129** 231-238.

41   Gerard JF & Gilles R (1972) The free amino-acid pool in *Callinectes sapidus* (Rathbun) tissues and its rôle in the osmotic intracellular regulation. *Journal of Experimental Marine Biology and Ecology* **10** 125-136.

42   Glashenn JS & Hand SC (1989) Metabolic heat dissipation and internal solute levels of *Artemia* embryos during changes in cell-associated water. *Journal of Experimental Biology* **145** 263-282.

43   Goethard JWC & Heinsius HW (1892) Biologie von *Noctiluca miliaris. Tidschriefter van de Nederlandsche Dierkundige Vereeniging* **2** 36.

44   Gordon MS & Tucker VA (1965) Osmotic regulation in the tadpoles of the crab-eating frog (*Rana cancrivora). Journal of Experimental Biology* **42** 437-445.

45   Gordon MS, Schmidt-Nielsen K & HM Kelly (1961) Osmotic regulation in the crab-eating frog (*Rana cancrivora). Journal of Experimental Biology* **38** 659-678.

46   Greenaway P (1988) Ion and water balance. *Biology of the Land Crabs* pp 211-248. Eds WW Burggren & BR McMahon. Cambridge: Cambridge University Press.

47   Griffith RW (1985) Habitat, phylogeny and the evolution of osmoregulatory strategies in primitive fishes. *Evolutionary Biology of Primitive Fishes* pp 69-80. Eds R E Foreman, A Gorbman, J M Dodd & R Olsson. New York: Plenum Press.

48 Griffith RW (1987) Freshwater or marine origin of vertebrates? *Comparative Biochemistry and Physiology* **87A** 523-531.

49 Griffith RW (1994) The life of the first vertebrate. *Bioscience* **44** 408-417.

50 Griffith RW & Pang PKT (1979) Mechanisms of osmoregulation in the coelacanth: evolutionary implications. *Occasional Papers of the Californian Academy of Sciences* **134** 79-93.

51 Griffith RW, Umminger BL, Grant BF, Pang PKT & Pickford GE (1974) Serum composition of the coelacanth, *Latimeria chalumnae* Smith. *Journal of Experimental Zoology* **187** 87-102.

52 Griffiths M (1968) *Echidnas*. Oxford: Pergamon Press.

53 Gross F & Zeuthen E (1948) The buoyancy of plankton diatoms: a problem of cell physiology. *Proceedings of the Royal Society of London B* **135** 383-389.

54 Grundy JE & Storey KB (1994) Urea and salt effects on enzymes from estivating and non-estivating amphibians. *Molecular and Cellular Biochemistry* **131** 9-17.

55 Hazelwood DH, Potts WTW & Fleming WR (1970) Further studies on the sodium and water metabolism of the freshwater medusa, *Craspedacusta sowerbyi*. *Zeitschrifte für vergleichende Physiologie* **67** 186-191.

56 Hillman PE & Pough FH (1976) Salt excretion in a beach lizard (*Ameiva quadrilineata*, Teiidae). *Journal of Comparative Physiology* **109** 169-175.

57 Hochachka PW & Somero GN (1984) *Biochemical Adaptation*. Princeton: Princeton University Press.

58 Holmes WN & Donaldson EM (1969) The body compartments and the distribution of electrolytes. *Fish Physiology. Vol 1. Excretion, Ionic Regulation, and Metabolism*, pp 1-89. Eds WS Hoar & DJ Randall. New York: Academic Press.

59 Jones MEE, Bradshaw SD, Fergusson B & Watts R (1990) Effect of available surface water on levels of antidiuretic hormone (lysine vasopressin) and water and electrolyte metabolism of the Rottnest Island quokka (*Setonix brachyurus*). *General and Comparative Endocrinology* **77** 75-87.

60 de Jorge FB, Petersen JA, Ditadi ASF & Sawaya P (1966) Biochemical studies in a eunicid (polychaete) of the littoral of São Paulo, Brazil. *Comparative Biochemistry and Physiology* **17** 535-551.

61 Kahn N & Swift E (1978) Positive buoyancy through ionic control in the nonmotile marine dinoflagellate *Pyrocystis noctiluca* Murray ex Schuett. *Limnology and Oceanography* **23** 649-658.

62 Katz U & Gabbay S (1986) Water retention and plasma and urine composition in toads (*Bufo viridis* Laur.) under burrowing conditions. *Journal of Comparative Physiology* **156** 735-740.

63 Katz U & Hanke W (1993) Mechanisms of hyperosmotic acclimation in *Xenopus laevis* (salt, urea or mannitol). *Journal of Comparative Physiology* **163B** 189-195.

64 Kennett JP (1982) *Marine Geology*. Englewood Cliffs: Prentice-Hall.

65 Krogh A (1939) *Osmotic Regulation in Aquatic Animals*. Cambridge: Cambridge University Press.

66 Lambert CC & Lambert G (1978) Tunicate eggs utilize ammonium ions for flotation. *Science* **200** 64-65.

67 Lien Y-HH, Pacelli MM & Braun EJ (1993) Characterization of organic osmolytes in avian renal medulla: a nonurea osmotic gradient system. *American Journal of Physiology* **264** R1045-R1049.

68 Loveridge JP & Withers PC (1981) Metabolism and water balance of active and cocooned African bullfrogs Pyxicephalus adspersus. *Physiological Zoölogy* **54** 203-214.

69   Lovett DL & Watts SA (1995) Changes in polyamine levels in response to acclimation salinity in gills of the blue crab *Callinectes sapidus* Rathbun. *Comparative Biochemistry and Physiology* **110B** 115-119.

70   Low PS (1985) Molecular basis of the biological compatibility of nature's osmolytes. *Transport Processes, Iono- and Osmoregulation*, pp 469-477. Eds R Gilles & M Gilles-Baillien. Berlin: Springer Verlag.

71   Macallum AB (1903) On the inorganic composition of the medusae *Aurelia flavidula* and *Cyanea arctica. Journal of Physiology* **29** 213-241.

72   Mackenzie FT (1975) Sedimentary cycling and the evolution of seawater. *Chemical Oceanography*, pp 309-364. Eds JP Riley & G Skirrow. London: Academic Press.

73   Mallatt J, Conley DM & Ridgeway RL (1987) Why do hagfish have gill 'chloride' cells when they need not regulate plasma NaCl concentration? *Canadian Journal of Zoology* **65** 1956-1965.

74   Mantel LH & Farmer LL (1983) Osmotic and ionic regulation. *The Biology of Crustacea*, pp 53-161. Ed LH Mantel. New York: Academic Press.

75   McClanahan LL (1972) Changes in the body fluids of burrowed spadefoot toads as a function of soil water potential. *Copeia* **1972** 209-216.

76   McInerney JE (1974) Renal sodium reabsorption in the hagfish, *Eptratretus stoutii. Comparative Biochemistry and Physiology* **49A** 273-280.

77   Minnich JE (1982) The use of water. *Biology of the Reptilia Volume 12 Physiology C Physiological Ecology* pp 325-395. Eds C Gans & FH Pough. London: Academic Press.

78   Mommsen TP & Walsh PJ (1989). Evolution of urea synthesis in vertebrates: the piscine connection. *Science* **243** 72-75.

79   Mommsen TP & Walsh PJ (1992) Biochemical and environmental perspectives on nitrogen metabolism in fishes. *Experientia* **48** 583-593.

80   Oglesby LC (1978) Salt and water balance. *Physiology of Annelids*, pp 555-658. Ed PJ Mill. London: Academic Press.

81   Parry G (1953) Osmotic and ionic regulation in the isopod crustacean *Ligia oceanica* (L.). *Journal of Experimental Biology* **30** 567-574.

82   Pierce SK & Minasian LL (1974) Water balance of a euryhaline sea anemone, *Diadumene leucolena. Comparative Biochemistry and Physiology* **49A** 159-167.

83   Pierce SK, Edwards SC, Mazzocchi PH, Klingler LJ & Warren MK (1984) Proline betaine: a unique osmolyte in an extremely euryhaline osmoconformer. *Biological Bulletin* **167** 495-500.

84   Pierce SK, Rowland-Faux LM & Crombie BN (1995) The mechanism of glycine betaine regulation in response to hyperosmotic stress in oyster mitochondria: a comparative study of Atlantic and Chesapeake Bay oysters. *Journal of Experimental Zoology* **271** 161-170.

85   Potts WTW (1954) The inorganic composition of the blood of *Mytilus edulis* and *Anodonta cygnea. Journal of Experimental Biology* **31** 376-385.

86   Potts WTW & Parry G (1964) *Osmotic and Ionic Regulation in Animals*. Oxford: Pergamon Press.

87   Potts WTW & Parry G (1964) Sodium and chloride balance in the prawn *Palaemonetes varians. Journal of Experimental Biology* **41** 591-601.

88   Prosser CL (1973) *Comparative Animal Physiology*. Philadelphia: Saunders College.

89   Randall DJ, Wood CM, Perry SF, Bergman H, Maloiy GMO, Mommsen TP & Wright PA (1989) Urea excretion as a strategy for survival in a fish living in a very alkaline environment. *Nature* **337** 165-166.

90   Robertson JD (1964) Osmotic and ionic regulation. *Physiology of the Mollusca*, pp 283-311. Eds KM Wilbur & CM Yonge. New York: Academic Press.

91   Robertson JD (1970) Osmotic and ionic regulation in the horseshoe crab *Limulus polyphemus* (Linnaeus). *Biological Bulletin* **138** 157-183.

92   Robertson JD (1976) Chemical composition of the body fluids and muscle of the hagfish *Myxine glutinosa* and the rabbit-fish *Chimaera monstrosa*. *Journal of Zoology* **178** 261-277.

93   Robson EA, Lockwood APM & Ralph R (1966) Composition of the blood in Onycophora. *Nature* **209** 533.

94   Rompsert AP (1976) Osmoregulation of the African clawed frog, *Xenopus laevis*, in hypersaline media. *Comparative Biochemistry and Physiology* **54A** 207-210.

95   Rozemeijer MJC & Plaut I (1993) Regulation of nitrogen excretion of the amphibious Blenniidae *Alticus kirki* (Guenther, 1868) during emersion and immersion. *Comparative Biochemistry and Physiology* **104A** 57-62.

96   Ruppert EE & Barnes RD (1994) *Invertebrate Zoology*. Fort Worth: Saunders College Publishing.

97   Sanders NK & Childress JJ (1988) Ion replacement as a buoyancy mechanism in a pelagic deep-sea crustacean. *Journal of Experimental Biology* **138** 333-343.

98   Sasayama Y, Oguro C, Ogasawara T, Hirano T, Harumi T & Wisessang S (1990) The crab-eating frog *Rana cancrivora*: serum mineral concentrations and histology of the ultimobranchial gland and the parathyroid gland. *General and Comparative Endocrinology* **80** 399-406.

99   Seck C (1957) Untersuchungen zur frage der ionenregulation bei in brackwasser lebenden evertebraten. *Kieler Meereforschungen* **13** 220-243.

100   Shoemaker VH & McClanahan LL (1975) Evaporative water loss, nitrogen excretion and osmoregulation in phyllomedusine frogs. *Journal of Comparative Physiology* **100** 331-345.

101   Shoemaker VH, Hillman SS, Hillyard SD, Jackson DC, McClanahan LL, Withers PC & Wygoda ML (1992) Exchange of water, ions, and respiratory gases in terrestrial amphibians. *Environmental Physiology of the Amphibians*, pp 125-150. Eds ME Feder & WW Burggren. Chicago: University of Chicago Press.

102   Simpson JW, Allen K & Awapana J (1959) Free amino acids in some aquatic invertebrates. *Biological Bulletin* **117** 371-378.

103   Skadhauge E, Clemens E & Maloiy GMO (1980) The effects of dehydration on electrolyte concentrations and water content along the large intestine of a small ruminant: the dik-dik antelope. *Journal of Comparative Physiology* **135** 165-173.

104   Smith HW (1930) Metabolism of the lungfish, *Protopterus æthiopicus*. *Journal of Biological Chemistry* **88** 97-130.

105   Smith HW (1932) Water regulation and its evolution in fishes. *Quarterly Reviews of Biology* **7** 1-26.

106   Smith HW (1936) The retention and physiological role of urea in the Elasmobranchii. *Biological Reviews* **11** 49-82.

107   Spaargaren DH (1979) A comparison of the composition of physiological saline solutions with that of calculated pre-cambrian seawater. *Comparative Biochemistry and Physiology* **63A** 319-323.

108   Sutcliffe DW (1963) The chemical composition of haemolymph in insects and some other arthropods, in relation to their phylogeny. *Comparative Biochemistry and Physiology* **9** 121-135.

109   Taplin LE, Grigg GC & Beard L (1993) Osmoregulation of the Australian freshwater crocodile, *Crocodylus johnstoni*, in fresh and saline waters. *Journal of Comparative Physiology* **163B** 70-77.

110 Thompson GG & Withers PC (1992) Osmoregulatory adjustments by three atherinids (*Leptatherina presbyteroides*; *Craterocephalus mugiloides*; *Leptatherina wallacei*) to a range of salinities. *Comparative Biochemistry and Physiology* **103A** 725-728.

111 Thorson TB, Cowan CM & Watson DE (1967) *Potamotrygon* spp.: elasmobranchs with low urea content. *Science* **158** 375-377.

112 Treherne JE & Maddrell SHP (1967) Membrane potentials in the central nervous system of a phytophagous insect (*Carausius morosus*). *Journal of Experimental Biology* **46** 413-421.

113 Urist MR (1962) The bone-body continuum: calcium and phosphorus in the skeleton of extinct and living vertebrates. *Perspectives in Biology and Medicine* **6** 75-115.

114 Warburg MR (1993) *Evolutionary Biology of Land Isopods*. Berlin: Springer Verlag.

115 Warren MK & Pierce SK (1982) Two cell volume regulatory systems in the *Limulus* myocardium: an interaction of ions and quaternary ammonium compounds. *Biological Bulletin* **163** 504-516.

116 Wilkinson L (1990) *Statistics*. Evanston: Systat Inc.

117 Withers PC (1992) *Comparative Animal Physiology*. Philadelphia: Saunders College Publishing.

118 Withers PC (1993) Metabolic depression during aestivation in the Australian frogs, *Neobatrachus* and *Cyclorana*. *Australian Journal of Zoology* **41** 467-473.

119 Withers PC, Morrison G & Guppy M (1994) Buoyancy role of urea and TMAO in an elasmobranch fish, the Port Jackson shark, *Heterodontus portusjacksoni*. *Physiological Zoölogy* **67** 693-705.

120 Withers PC, Morrison G, Hefter GT & Pang T-S (1994) Role of urea and methylamines in buoyancy of elasmobranchs. *Journal of Experimental Biology* **188** 175-189.

121 Withers PC & Guppy M (1996) Do Australian desert frogs coaccumulate counteracting solutes in urea during aestivation? *Journal of Experimental Biology* (In Press).

122 Wray S & Wilkie DR (1995) The relationship between plasma urea levels and some muscle trimethylamine levels in *Xenopus laevis*: a ^{31}P and ^{14}N nuclear magnetic resonance study. *Journal of Experimental Biology* **198** 373-378.

123 Wright PA (1995) Nitrogen excretion: three end products, many physiological functions. *Journal of Experimental Biology* **198** 273-281.

124 Wright PA, Felskie A & Anderson PM (1995) Induction of ornithine-urea cycle enzymes and nitrogen metabolism and excretion in rainbow trout (*Oncorhynchus mykiss*) during early life stages. *Journal of Experimental Biology* **198** 127-135.

125 Yancey PH (1988) Osmotic effectors in kidneys of xeric and mesic rodents: corticomedullary distributions and changes with water availability. *Journal of Comparative Physiology* **158B** 369-380.

126 Yancey PH & Somero GN (1979) Counteraction of urea destabilization of protein structure by methylamine osmoregulatory compounds of elasmobranch fishes. *Biochemical Journal* **183** 317-323.

127 Yancey PH, Clark ME, Hand SC, Bowlus RD & Somero GN (1982) Living with water stress: evolution of osmolyte systems. *Science* **217** 1214-1222.

128 Zatta P (1987) The relationship between plasma proteins and intracellular free amino acids during osmotic regulation in *Carcinus maenas*. *Journal of Experimental Zoology* **242** 131-136.

The Comparative Endocrinology of Calcium Regulation
Eds C Dacke, J Danks, I Caple & G Flik, pp 43-53
Journal of Endocrinology Ltd, Bristol (1996)

Calcium homeostasis in fish

G Flik and P M Verbost[1]

Department of Animal Physiology, Faculty of Science, University of
Nijmegen, Toernooiveld, 6525ED Nijmegen, The Netherlands; [1]Organon
Int., Molenstraat 110, PO Box 20, 5340 BH Oss, The Netherlands

The importance of gills in calcium homeostasis

Fish control plasma calcium levels within reasonably narrow limits (19, 23, 49, 50) and typical plasma calcium concentrations are around 3 mmol/l of which roughly 50% is in the ionic form Ca^{2+} (23); much higher levels of calcium are found in plasma of female fish during ovarian maturation and this extra calcium is bound to vitellogenin, the calcium *activity* in the plasma is kept around 1.5 mmol/l. It thus appears that fish too aim at calcium homeostasis, although the deviations around a set-point level of plasma calcium in fish are far larger than observed in mammals (23, 50). Fish take up calcium from their environment, the water, via their gills and this uptake of Ca^{2+} suffices to guarantee calcium homeostasis and growth (13, 35). Branchial Ca^{2+} uptake is an active (35) and a more or less continuous process and largely independent of the ambient calcium concentration. The branchial epithelium is exposed to solutions containing less than 0.1 mmol/l Ca in soft fresh water and over 10 mmol/l Ca in seawater. The electrochemical conditions for branchial Ca^{2+} flow in fresh water and seawater have recently been discussed (20): the driving force for passive movement of Ca^{2+} over the epithelium is directed outwardly both in fresh water and seawater, mainly as a result of the change in the transepithelial potential that shifts from slightly negative (typical value -2 mV) in freshwater to around 25 mV positive in seawater fish. Yet, both freshwater and seawater tilapia have a significant Ca^{2+} influx, for which the fish requires active transport mechanisms.

Many fish appear to live and grow normally when fed a calcium deficient diet. Therefore, the role of the intestine in calcium uptake by (freshwater) fish is often considered accessory. As branchial Ca^{2+} uptake is adjusted when the diet is deficient in calcium, one would conclude that intestinal calcium handling is not so important for calcium homeostasis; however, it may play an important, as yet enigmatic role in e.g. phosphate or magnesium handling. Highly significant mucosal-to-serosal and active transport occurs in isolated fish proximal intestine (17, 39). Ca^{2+} influx over this epithelium is Na^+ dependent as measured in an Ussing chamber set-up. The rate of ATP-driven Ca^{2+} transport in the intestinal basolateral plasma membrane vesicles is very low: too low to explain the Ca^{2+} fluxes that occur in this epithelium. However, in the tilapia enterocyte basolateral plasma membrane a powerful Na^+/Ca^{2+} exchange activity may fulfil this function, which corroborates the Na^+ dependence in the Ussing chamber experiments (17). Once adapted to seawater, it appears that the intestine of tilapia has remarkably changed. The seawater fish drinks water containing around 10 mmol/l Ca. *In vitro,* a net Ca^{2+} influx was no longer measured in the intestine of seawater fish. The Na^+/K^+-ATPase activity in the basolateral membrane (BLM) had increased in line with a water transport function (solute-linked water transport) for the intestine in seawater. Coincidently, the activities of the Ca^{2+}-ATPase and, more pronouncedly, of the Na^+/Ca^{2+}-exchanger showed very significant decreases in seawater fish compared with freshwater fish. It thus appears that a physiological adaptation evokes an adjustment of Ca^{2+} carrier densities in the basolateral plasma membrane compartment, where the extrusion step of transepithelial Ca^{2+} transport is realized (39). Only a few studies on intestinal Ca^{2+} transport in fish are found in the literature, and this is surprising when one considers the variety in types of intestine found in fish and the powerful models they may represent. Recently (11, 19), we have concluded that, in tilapia, intestinal secretion of calcium may be pivotal in calcium homeostasis. The endocrines and their target mechanisms that govern calcium transport in fish intestine include the vitamin-D system (see Sundell *et al.*, this volume). Fish must rely on intestinal uptake of phosphate and magnesium, as these ions are not taken up via the gills (18). The intestine of seawater fish is specialized to take up water to compensate for osmotic losses of water via its integument. The intestinal epithelium faces on its mucosal side calcium concentrations up to 25 mmol/l (as measured in tilapia; 28) and these concentrations may be anticipated to be higher in fish that eat bones, shells or corals.

The kidney of freshwater fish produces a dilute urine to counteract osmotic inflow of water via skin and gills (8, 38); calcium loss is minimized through active reabsorption of ultrafiltered calcium (2, 34, 37, 38). In seawater, fish reduce urine flow (26) but secrete calcium (and magnesium and phosphates; 37). Calcium concentrations in the urine of seawater fish always exceed those in the plasma. Nephric epithelium at its mucosal side is exposed to solutions containing ultrafiltered, millimolar concentrations of calcium. In fish adapted to seawater, renal Ca^{2+} reabsorption is decreased (2, 8, 37). Also, renal membrane preparations of several euryhaline species exhibited lower Na^+-pump activity when the fish was exposed to seawater (7, 43). The hypothesis was formulated that in the kidney of seawater fish less transcellular Ca^{2+} transport is required and that this should be reflected by a decrease in Ca^{2+} extrusion activity. Two observations allow us to conclude that a Ca^{2+}-ATPase in the basolateral plasma membrane of renal cells is involved in Ca^{2+} reabsorption. First, we cannot demonstrate Na^+/Ca^{2+}-exchange activity in tilapia renal membrane preparations (3). This observation is in line with studies suggesting that in fish renal Ca^{2+} reabsorption primarily depends on intracellular ATP and is largely Na^+-independent (37). Second, a high-affinity, high-capacity and ATP-driven Ca^{2+} pump activity decreases significantly upon seawater adaptation (4, 7). Proceeding from a homogeneous Ca^{2+}-ATPase activity (indicated by kinetic analyses), similar purification numbers and specific activities for Na^+/K^+-ATPase in renal membrane preparations of freshwater and seawater fish are found and we conclude that the relative density of a single class of

ATP-driven Ca^{2+} pumps decreases upon seawater adaptation. Assuming comparable requirements for cellular Ca^{2+} homeostasis in kidney cells of freshwater and seawater tilapia, a decreased need for Ca^{2+} reabsorption (depending on transcellular Ca^{2+} transport) appears to be reflected by a decreased Ca^{2+}-ATPase activity. Fish, again, and in particular their kidneys should be appreciated as versatile models for (comparative) renal physiology.

Active transport of Ca^{2+}

Cell-mediated Ca^{2+} transport implies that calcium ions have to cross two plasma membrane barriers, viz. the apical and the basolateral plasma membrane. This leaves two plasma membrane sites requiring different forms of regulation (12, 17, 32, 44). Assuming that the movement of calcium over biomembranes is not diffusive but mediated through carrier proteins, distinct and carrier-specific kinetic criteria were advanced for Ca^{2+} transporters in the different plasma membrane domains. For all three epithelia mentioned, *mucosal* concentrations of calcium are mostly in the millimolar range. *Cytosolic* Ca^{2+} concentrations (resting levels) are in the submicromolar range (typical value 100 nmol/l (5, 44)). We predict, therefore, that carriers in the apical membrane of a calcium transporting epithelium will have a half-maximum activation concentration of Ca^{2+} (K_m) in the millimolar range, in contrast to those in the basolateral plasma membrane with a K_m in the nanomolar range. As the interior of the cell is electrically negative relative to the exterior, it follows too that cellular Ca^{2+} influx is passive, down an electrochemical gradient, and that it is the Ca^{2+} extrusion to the serosal compartment that requires energy to overcome the steep electrochemical gradient. Ca^{2+} flux ratio analysis in intact trout and isolated head preparations of this fish has unequivocally demonstrated that active Ca^{2+} transport mechanisms must underlie branchial uptake of Ca^{2+} from the water (35).

In recent years a steady flow of papers has advanced evidence that Ca^{2+} influx at least in branchial and intestinal (and likely in renal) Ca^{2+} transporting cells is controlled by stanniocalcin (STC) (4, 18, 49). STC is a hormone produced by the so-called corpuscles of Stannius endocrine glands found only in holostean and teleostean fish species. In gills and intestine STC inhibits Ca^{2+} entry (16, 45) and it thus controls the permeability for Ca^{2+} of the apical membrane through a second messenger (cyclic AMP) dependent pathway, most likely a second messenger operated calcium channel (45). STC is a fast acting hormone providing the fish with a tool for the rapid control of Ca^{2+} flow. A higher STC metabolic clearance rate and secretion rate in seawater eel compared with that in freshwater eel (24) indicate that seawater may impose a greater calcium challenge to the animal than fresh water. Implicitly too, such results corroborate the notion of cell-mediated, transcellular and thus active calcium transport in seawater fish gills.

Studies on Ca^{2+} transport in isolated apical membranes of fish tissues are essentially restricted to intestinal brush border membranes. Apical membranes of branchial epithelium have not been isolated for this pupose. We have obtained good evidence for a carrier-mediated transport of Ca^{2+} in tilapia intestinal brush border vesicles (28). Although there are indications that intestinal Ca^{2+} transport is controlled by STC (11), we have failed so far to show an influence of STC (or a second messenger) on this particular transport. A consistent finding is that calcium transport in brush border vesicles, also in tilapia (Klaren, personal observation) is linked to a purinergic, P_2-type (suramin-sensitive) receptor, assumed to be a calcium channel (51). The possible link between STC and suramin-sensitive Ca^{2+} transport is presently under investigation in our laboratory.

Studies on active Ca^{2+} transport mechanisms in fish gills started in the 1970s with the search for an energized Ca^{2+} pump. At that time, a Ca^{2+} ATPase with a low affinity for Ca^{2+} (millimolar) was postulated to provide the driving force. Later on this so-called ATPase was defined as a *non-specific phosphatase* activity. Subsequently, attention was focused on the possible presence of a Ca^{2+} ATPase with characteristics compatible with a pump that extrudes Ca^{2+} from the ionocyte against the steep electrochemical gradient for Ca^{2+} maintained over the basolateral plasma membrane of these cells (14). Later (46), it was demonstrated that in addition to this high-affinity Ca^{2+} ATPase with characteristics of P-type ATPases (30) a Na^+/Ca^{2+}-exchanger should also be considered in Ca^{2+} extrusion. Ca^{2+} flux ratio analysis in intact trout and trout isolated head preparations has unequivocally demonstrated that active Ca^{2+} transport mechanisms must underlie branchial uptake of Ca^{2+} from the water (35).

In the complex epithelium of the gills' chloride cells, specialized ion transporting cells mediate Ca^{2+} uptake (14, 31, 36). A series of criteria were met to identify positively the calcium pumps in the basolateral plasma membrane fraction of the chloride cells. First, the location of a calcium pump in the basolateral plasma membrane of an ion transporting cell implies that the pump is co-located with markers for this membrane compartment. The Na^+/K^+-ATPase activity is such a marker as this enzyme is more abundant in the chloride cells with their extensive invaginations of the basolateral plasma membrane (12, 30). A membrane isolation procedure based on the purification of Na^+/K^+-ATPase has yielded membrane preparations enriched in basolateral plasma membranes and, in particular, plasma membranes of the ionocytes of the branchial epithelium. Procedures have been described (12) that allow the isolation of membrane fractions that are essentially devoid of membranes from blood cells, or subcellular membranes from endoplasmic reticulum, Golgi apparatus or mitochondria. This is a prerequisite, as Ca^{2+} pumps are present in all membranes facing the cytosol, and only the plasma membrane calcium pumps extrude Ca^{2+} for epithelial calcium uptake.

A specific problem encountered in the study of fish branchial plasma membranes is an often exuberant non-specific phosphatase activity: in a large pool of phosphatases one has to identify the Ca^{2+} transporting enzyme, that hydrolyses maybe less than 5% of the ATP hydrolysed in total. Specific ATP-preference, submicromolar affinity for Ca^{2+} and Michaelis-Menten kinetics of phosphatase activity are difficult to assess in such assays, but these criteria were met in the demonstration of a high-affinity Ca^{2+}-ATPase in tilapia, eel, trout, carp and fundulus gill plasma membranes (14, 16, 35). Yet, homogeneous kinetics, high affinity for Ca^{2+} and ATP dependence still do not guarantee the discrimination of an actual transporting ATPase. Rat liver plasma membrane contains two Ca^{2+}-ATPase activities with comparable kinetics, one of which is the molecular correlate of a calcium pump, the other is an ecto-ATPase involved in regulating extracellular levels of adenosine nucleotides (29). It follows that, in addition to copurification with Na^+/K^+-ATPase and proper Ca^{2+} kinetics, more criteria are required to identify a calcium pump involved in epithelial transport. This is realized in transport assays with membrane vesicles (13, 22, 33, 44).

Using resealed plasma membrane vesicles, and applying the same criteria as mentioned above for the ATPase assays, the activity of the calcium pump can be assayed as ATP-driven $^{45}Ca^{2+}$ accumulation into the vesicular space. This 'vectorial' assay unambiguously reveals the activity of a calcium pump. The plasma-membrane-associated calcium pump of gills proved to be calmodulin-dependent (35), a characteristic of plasma membrane calcium pumps (P-type Ca^{2+}-ATPases) in general (40). In the plasma membrane preparations used for these studies, thapsigargin (9, 42) did not inhibit Ca^{2+} pump activity in gill plasma membrane preparations or affect its kinetics, and this observation confirmed the purity of this preparation. It seems justified then to state that fish gill plasma membranes contain a P-type Ca^{2+}-ATPase, which could be the driving force for Ca^{2+} transport in the tissue.

The Na^+/Ca^{2+}-exchanger is the second mechanism for energized transport of Ca^{2+} across plasma membranes. Understanding Ca^{2+} extrusion in cells now requires the simultaneous analysis of ATP- and Na^+-dependent Ca^{2+} pumps. This Na^+-dependent carrier was first assumed to be of particular importance in excitable tissues, where Na^+ and Ca^{2+} countercurrents underly the events related to the generation of action potentials. However, the carrier has been demonstrated in non-excitable cells as well and thus it may function in Ca^{2+} transport in these cells. The first evidence for a Na^+/Ca^{2+} exchanger came from studies on tilapia enterocyte plasma membrane that proved to be a poor source of Ca^{2+}-ATPase, but contains an extremely active exchanger (15). The carrier exchanges three Na^+ for one Ca^{2+} and participates in Ca^{2+} extrusion in the tilapia enterocyte (17, 39), where Na^+/K^+-ATPase activity maintains an inward sodium gradient as driving force. A caveat concerns the extrapolation of such data to, for example, mammals since in rat renal and intestinal tissue no clear function in Ca^{2+} extrusion could be attributed to the exchanger (21, 25, 44). The exchanger was subsequently also demonstrated in gill epithelium (46), but appears essentially absent from renal tissue of tilapia (3).

The kinetic parameters of the exchanger were compared with those of the ATP-driven calcium pump and this biochemical comparison led us to conclude that in the gills the ATP-driven Ca^{2+} pump may play an equal or even more pronounced role than the exchanger in active Ca^{2+} transport in branchial epithelium: we calculated an activity of the ATPase that was almost twice that of the exchanger at prevailing cytosolic Ca^{2+} levels (46). Anticipating a higher turnover for Na^+ (30) in the gills of seawater fish compared with freshwater fish, a more pronounced role for the exchanger was predicted in calcium handling by seawater fish gills. However, kinetic analyses revealed that the relative densities of the two calcium transporters do not differ in tilapia well-acclimated to either fresh water or seawater. The involvement and role of the exchanger in calcium transport in branchial epithelium awaited physiological studies. Such studies have recently been realized using opercular membranes as a model for fish gills. Ca^{2+} transport in *Fundulus* opercular epithelium (which is extremely rich in chloride cells) is for the major part Na^+-dependent (47). This could indicate that the Ca^{2+}-ATPase in branchial and opercular tissues is more a house-keeping enzyme to guarantee cellular Ca^{2+} homeostasis than a transepithelial transport (vectorial) extrusion mechanism. Extrapolation of data on the killifish to the tilapia is as yet speculation; however, the presence of Ca^{2+}-ATPase and Na^+/Ca^{2+} exchange activity in *Fundulus* gills means that the mechanisms present in the gills of tilapia and killifish are comparable.

Calciotropic hormones in fish

The prolactin gene family products

Prolactin exerts hypercalcaemic actions in fish, it stimulates Ca^{2+} influx from the water and specifically enhances ATP-driven Ca^{2+} pump activity in the plasma membranes of branchial epithelium (1, 15, 16, 19, 41). It thus appears that the calcium pump capacity of the branchial epithelium is adjusted to the calcium uptake from the water under control of this hormone. At this moment we cannot exclude that increased Ca^{2+}-ATPase activity relates to an increased Ca^{2+} turnover in the epithelium and thus reflects a house-keeping task of the enzyme rather than a transepithelial transport function. In tilapia, homologous prolactin (41) increases the branchial ionocyte density in a dose-dependent manner (19) and, in parallel, the branchial Ca^{2+} influx. Therefore, we assume that in prolactin-treated fish new populations of chloride cells become active. Moreover, the density of the Ca^{2+}-ATPase relative to the Na^+/K^+-ATPase increases upon prolactin treatment, and this suggests that the expression of this calcium pump in fish gills is specifically enhanced by this pituitary hormone. Certainly, prolactin is not the only candidate for a hypercalcaemic calciotrope in fish. In trout, not prolactin but cortisol enhanced calcium pumping activity in the plasma membrane fraction of the gills (10). Another pituitary candidate for hypercalcaemic actions in trout is somatolactin (27). The hormone has so far only been found in fish and is closely related to prolactin and growth hormone. It is produced by the periodic acid Schiff's reagent (PAS)-positive (somatolactin is in most species glycosylated) cells of the pars intermedia of fish, which may be activated or inhibited by low and high environmental levels of calcium respectively (27). As the uptake of calcium from the water in some species (e.g. killifish, tilapia) in freshwater is under stimulatory control by prolactin (20), hypophysectomy results in hypocalcaemia (50), which may be overcome specifically by prolactin replacement. The calciotropic effects exerted by prolactin develop slowly (after two to four days) and their demonstration most likely can only be demonstrated in transporting cells newly developed under the altered prolactin tonus (19). To date nothing is known about calcitropic hormones with hypercalcaemic actions in seawater species. It is, however, tempting to speculate that other members of the prolactin gene family, viz. growth hormone and somatolactin, are somehow involved. Prolactin cells become inactivated by high external calcium levels as encountered in seawater (48). Reports on a possible role of growth hormone in calcium metabolism in seawater fish are to the best of our knowledge not available.

Stanniocalcin

The uptake of calcium is under a predominantly inhibitory control by the hormone STC (49). Removal of the corpuscles of Stannius, the source of STC, invariably results in hypercalcaemia, at least when the water contains calcium. The hypercalcaemia that develops in the absence of stanniocalcin is positively related with the calcium content of the water (23). More elaborate and detailed information on the role of STC in fish can be found in the chapter by Verbost and coworkers in this volume.

Two topics we would like to discuss here in relation to the action of STC in fish concern the possible targets and the existence of an endogenous antagonist. There is a relative wealth of information on the action of STC on the gills. The prevailing idea is that the hormone controls calcium entry through control of calcium channels in the apical membrane of the calcium transporting cell and this, by definition, must be a second messenger-operated calcium channel (SMOC). Assuming that in the principle targets of STC (gills, intestine and kidney tubules) the target mechanism, viz. a SMOC, is the same, the data available are well-explained. An inhibitory action of STC reduces mucosa to serosa calcium transport, which means that the uptake in gills and intestine and the re-uptake from the ultrafiltered plasma in the nephron are controlled. With an increased amount of STC being active in seawater fish, an answer is given to the more pronounced and imminent calcium overload in this high calcium environment at the level of the gills and intestine (fish drink seawater to compensate for the osmotic loss of water, but certainly do not need the calcium in the water drunk); also reduced reabsorption in the nephron in seawater fish is compatible with enhanced STC activity. Unfortunately, very little information is available on the effects of STC on intestine or kidney calcium handling.

The second topic concerns the question of whether we should postulate an antagonist for STC. Hormones do act in tandem (parathyroid hormone and calcitonin, insulin and glucagon) and require feedback to fine-regulate a process. When one considers that it takes only minutes to get response to STC and that its half-life in the circulation is around 2 h (24), it follows that the STC signal must be counteracted. This could be through either autoregulatory phenomena at the levels of the SMOC (Ca^{2+}-dependent or cyclic-nucleotide-dependent phosphorylation/dephosphorylation) or through the action of an antagonist. The work of Danks and coworkers (6) suggests that the corpuscles of Stannius of trout eel and salmon may contain parathyroid hormone related protein (PTHrP) and this could be a candidate for an endogenous STC antagonist. As the removal of the corpuscles of Stannius always results in hypercalcaemia and not in hypocalcemia, it would follow that an antagonistic effect of PTHrP is either mild or relates, for instance, to the autocrine control of STC release. We are at present carrying out experiments relating to this topic.

Acknowledgements

GF was supported by a travel grant from the Royal Netherlands Academy of Arts and Sciences (KNAW); PMV was associated with the department as KNAW-fellow until November 1995.

References

1   Ayson FG, Kaneko T, Tagawa M, Hasegawa S, Grau EG, Nishioka RS, King DS, Bern HA & Hirano T (1993) Effects of acclimation to hypertonic environment on plasma and pituitary levels of two prolactins and growth hormone in two species of tilapia, *Oreochromis mossambicus* and *Oreochromis niloticus*. *General Comparative Endocrinology* **89** 138-148.

2   Björnsson BTh & Nilsson S (1985) Renal and extra-renal excretion of calcium in the marine teleost, *Gadus morhua*. *American Journal of Comparative Physiology* **248** R18-R22.

3   Bijvelds MJC, van der Heijden AJH, Flik G, Verbost PM, Kolar ZI & Wendelaar Bonga SE (1995) Calcium pump activities in the kidneys of *Oreochromis mossambicus*. *Journal of Experimental Biology* **198** 1351-1357.

4   Butler DG (1993) Stanniectomy increases renal magnesium and calcium excretion in freshwater North American eels (*Anguilla rostrata*). *Journal of Experimental Biology* **181** 107-118.

5   Carafoli E (1987) Intracellular calcium homeostasis. *Annual Review of Biochemistry* **56** 395-433.

6   Danks JA, Devlin AJ, Ho PM, Diefenbach-Jagger H, Power DM, Canario A, Martin TJ & Ingleton PM (1993) Parathyroid hormone related protein is a factor in normal fish pituitary. *General Comparative Endocrinology* **92** 201-212.

7   Doneen BA (1993) High-affinity Ca^{2+}-Mg^{2+}-ATPase in kidney of euryhaline. *Gillichthys mirabilis*: kinetics, subcellular distribution and effects of salinity. *Comparative Biochemistry and Physiology* **106B** 719-728.

8   Elger E, Elger B, Hentschel H & Stolte H (1987) Adaptation of renal function to hypotonic medium in the winter flounder (*Pseudopleuronectes americanus*). *Journal of Comparative Physiology B* **157** 21-30.

9   Favero TG & Abramson JJ (1994) Thapsigargin-induced Ca^{2+} release from sarcoplasmic reticulum and asolectin vesicles. *Cell Calcium* **15** 183-189.

10   Flik G & Perry SF (1989) Cortisol stimulates whole body calcium uptake and the branchial calcium pump in freshwater rainbow trout. *Journal of Endocrinology* **120** 75-82.

11   Flik G & Verbost PM (1993) Calcium transport in fish gills and intestine. *Journal of Experimental Biology* **184** 17-29.

12   Flik G & Verbost PM (1994) Ca^{2+} transport across plasma membranes In *Biochemistry and Molecular Biology of Fishes*, vol 3, pp 625-637. Eds PW Hochachka & TP Mommsen. Amsterdam: Elsevier.

13   Flik G, Fenwick JC, Kolar Z, Mayer-Gostan N & Wendelaar Bonga SE (1985) Whole-body calcium flux rates in cichlid teleost fish, *Oreochromis mossambicus* adapted to freshwater. *American Journal of Comparative Physiology* **249** R432-R437.

14   Flik G, van Rijs JH & Wendelaar Bonga SE (1985) Evidence for high-affinity
      Ca^{2+}-ATPase activity and ATP driven Ca^{2+}-transport in membrane preparations of the gill
      epithelium of the cichlid fish Oreochromis mossambicus. *Journal of Experimental
      Biology* **119** 335-347.

15   Flik, G, Fenwick JC, Kolar Z, Mayer-Gostan N & Wendelaar Bonga SE (1986) Effects of
      ovine prolactin on calcium uptake and distribution in *Oreochromis mossmabicus*.
      American Journal of Comparative Physiology **250** R161-R166.

16   Flik G, Fenwick JC & Wendelaar Bonga SE (1989) Calcitropic actions of prolactin in
      freshwater North American eel (*Anguilla rostrata* LeSueur) *American Journal of
      Comparative Physiology* **257** R74-R79.

17   Flik G, Schoenmakers ThJM, Groot JA, van Os CH & Wendelaar Bonga SE (1990)
      Calcium absorption by fish intestine: the involvement of ATP-and sodium-dependent
      calcium extrusion mechanisms. *Journal of Membrane Biology* **113** 13-22.

18   Flik G, van der Velden JA, Dechering KJ, Verbost PM, Schoenmakers ThJM, Kolar ZI &
      Wendelaar Bonga SE (1993) Ca^{2+} and Mg^{2+} transport in gills and gut of tilapia,
      Oreochromis mossambicus: a review. *Journal of Experimental Zoology* **265** 356-366.

19   Flik G, Rentier-Delrue F & Wendelaar Bonga SE (1994) Calcitropic effects of
      recombinant prolactins in *Oreochromis mossambicus*. *American Journal of Comparative
      Physiology* **266** R1302-R1308.

20   Flik G, Klaren PHM, Schoenmakers ThJM, Bijvelds MJC, Verbost PM & Wendelaar
      Bonga SE (1996) Cellular calcium transport in fish: unique and universal mechanisms.
      Physiology and Zoology (In Press).

21   Friedman PA & Gesek FA (1993) Calcium transport in renal epithelial cells. *American
      Journal of Comparative Physiology* **264** F181-F198.

22   Ghijssen WEJM, de Jong MD & van Os CH (1982) ATP-dependent calcium transport and
      its correlation with Ca^{2+}ATPase activity in basolateral membranes of rat duodenum.
      Biochimica et Biophysica Acta **689** 327-336.

23   Hanssen RJGM, Lafeber FPJG, Flik G & Wendelaar Bonga SE (1989) Ionic and total
      calcium levels in the blood of the european eel (*Anguilla anguilla*): effects of
      stanniectomy and stanniocalcin replacement therapy. *Journal of Experimental Biology*
      141 177-186.

24   Hanssen RGJM, Mayer-Gostan N, Flik G & Wendelaar Bonga SE (1993) Stanniocalcin
      kinetics in freshwater and seawater european eel (*Anguilla anguilla*). *Fish Physiology and
      Biochemstry* **10** 491-496.

25   Heeswijk MPE, Geertsen JAM & van Os CH (1984) Kinetic properties of the
      ATP-dependent Ca^{2+} pump and the Na^+/Ca^{2+} exchange system in basolateral membranes
      from rat kidney cortex. *Journal of Membrane Biology* **79** 19-31.

26   Hickman CP Jr (1968) Glomerular filtration and urine flow in the euryhaline southern
      flounder, *Paralichthys lethostigma*, in seawater. *Canadian Journal of Zoology* **46**
      427-437.

27   Kaneko T & Hirano T (1993) Role of prolactin and somatolactin in calcium regulation in
      fish. *Journal of Experimental Biology* **184** 31-45.

28   Klaren PHM, Flik G, Lock RAC & Wendelaar Bonga SE (1993) Ca^{2+} Transport across
      intestinal brushborder membranes of the cichlid teleost. *Oreochromis mossambicus*.
      Journal of Membrane Biology **132** 157-166.

29   Lin SH & Russell WE (1988) Two Ca^{2+}-dependent ATPases in rat liver plasma
      membrane. *Journal of Biological Chemistry* **263** 12253-12258.

30   Maetz J & Bornancin M (1975) Biochemical and biophysical aspects of salt extrusion by chloride cells in teleosts. *Fortschritt Zoology* **23** 322-362.

31   McCormick SD, Hasegawa S & Hirano T (1992) Calcium uptake in the skin of a freshwater teleost. *Proceedings of the National Academy of Sciences of the USA* **89** 3635-3638.

32   Mircheff AK & Wright EM (1976) Analytical isolation of plasma membranes of intestinal epithelial cells: identification of Na,K-ATPase rich membranes and the distribution of enzyme activities. *Journal of Membrane Biology* **28** 309-333.

33   Murer H & Gmaj P (1986) Transport studies in plasma membrane vesicles isolated from renal cortex. *Kidney International* **30** 171-186.

34   Nishimura H & Imai M (1982) Control of renal function in freshwater and marine teleosts. *Federation Proceedings* **41** 2355-2360.

35   Perry SF & Flik G (1988) Characterization of branchial transepithelial calcium fluxes in freshwater trout, *Salmo gairdnerii. American Journal of Comparative Physiology* **254** R491-R498.

36   Perry SF, Goss CG & Fenwick JC (1992) Interrelationship between gill chloride cell morphology and calcium uptake in freshwater teleosts. *Fish Physiological Biochemistry* **10** 327-337.

37   Renfro JL, Dickman KG & Miller DS (1982) Effect of Na$^+$ and ATP on peritubular Ca transport by the marine teleost renal tubule. *American Journal of Comparative Physiology* **243** R34-R41.

38   Schmidt-Nielsen B & Renfro JL (1975) Kidney function of the American eel, *Anguilla rostrata. American Journal of Comparative Physiology* **228** 420-43.

39   Schoenmakers ThJM, Verbost PM, Flik G & Wendelaar Bonga SE (1993) Transcellular intestinal calcium transport in freshwater and seawater fish and its dependence on sodium/calcium exchange. *Journal of Experimental Biology* **176** 195-206.

40   Strehler EE (1991) Recent advances in the molecular characterization of plasma membrane Ca^{2+} pumps. *Journal of Membrane Biology* **120** 1-15.

41   Swennen D, Rentier-Delrue F, Auperin B, Prunet P, Flik G, Wendelaar Bonga SE, Lion M & Martial JA (1991) Production and purification of biologically active recombinant tilapia (*Oreochromis niloticus*) prolactins. *Journal of Endocrinology* **131** 219-227.

42   Thastrup O, Cullen PJ, Drobak BK, Hanley MR & Dawson AP (1990) Thapsigargin, a tumor promoter, discharges intracellular Ca^{2+} stores by specific inhibition of the endoplasmic reticulum Ca^{2+}-ATPase. *Proceedings of the National Academy of Sciences of the USA* **87** 2466-2470.

43   Trombetti F, Ventrella V, Pagliari A, Trigari G & Borgatti AR (1990) Mg^{2+}-dependent (Na$^+$+K$^+$)- and Na$^+$-ATPases in the kidneys of the gilthead bream (*Sparus auratus* L). *Comparative Biochemistry and Physiology* **97B** 343-354.

44   Van Os CH, van den Broek LAW, van Corven EJJM, Timmermans JAH & Dirven H (1988) Calcium homeostasis of epithelial cells. *Comparative Biochemistry and Physiology* **90A** 767-770.

45   Verbost PM, Flik G, Fenwick JC, Greco AM, Pang PKT & Wendelaar Bonga SE (1993) Branchial calcium uptake: possible mechanisms of control by stanniocalcin. *Fish Physiology and Biochemistry* **11** 205-215.

46   Verbost PM, Schoenmakers ThJM, Flik G & Wendelaar Bonga SE (1994) Kinetics of ATP- and Na$^+$-gradient driven Ca^{2+} transport in basolateral membranes from gills of freshwater and seawater adapted tilapia. *Journal of Experimental Biology* **186** 95-108.

47 Verbost PM, Bryson SE & Marshall WS (1995) Na^+ dependent Ca^{2+} uptake in isolated opercular epithelium of *Fundulus heteroclitus*. *Journal of Comparative Physiology B* (In Press).

48 Wendelaar Bonga SE, Löwik CJM & van der Meij JCA (1983) Effects of external Mg^{2+} and Ca^{2+} on branchial osmotic water permeability and prolactin secretion in the teleost fish *Sarotherodon mossambicus*. *General Comparative Endocrinology* **52** 222-231.

49 Wendelaar Bonga SE & Pang PKT (1986) Stannius corpuscles. In *Vertebrate Endocrinology: Fundamentals and Biomedical Implications*, pp 105-137. Eds PKT Pang & MP Schreibman. New York: Academic Press.

50 Wendelaar Bonga SE & Pang PKT (1991) Control of calcium regulating hormones in the vertebrates: parathyroid hormone, calcitonin, prolactin and somatolactin. *Internal Review of Cytology* **128** 139-213.

51 Wiley JS, Chen R & Jamieson GP (1993) The ATP^{4-} receptor-operated channel (P_{2z} class) of human lymphocytes allows Ba^{2+} and ethidium$^+$ uptake: inhibition of fluxes by suramin. *Archives of Biochemistry and Biophysics* **305** 54-60.

The Comparative Endocrinology of Calcium Regulation
Eds C Dacke, J Danks, I Caple & G Flik, pp 55-69
Journal of Endocrinology Ltd, Bristol (1996)

Cyclic adenosine monophosphate, a second messenger for stanniocalcin in tilapia gills?

P M Verbost, J C Fenwick[1], G Flik and S E Wendelaar Bonga

Department of Animal Physiology, Faculty of Science, University of
Nijmegen, Toernooiveld, 6525 ED Nijmegen, The Netherlands and
[1]Department of Biology, University of Ottawa, 30 George Glinski, Ottawa,
Canada K1N 6N5

Introduction

Cell-mediated branchial Ca^{2+} transport is under primary and inhibitory control by the fish hormone stanniocalcin (STC (34, 35)). STC inhibits Ca^{2+} influx within minutes and is thus considered a fast-acting hormone. In a trout isolated head preparation STC was shown to inhibit Ca^{2+} influx within 15 min (14). STC, a glycoprotein (2, 9, 15, 33) is assumed to bind to a membrane receptor on the chloride cell to influence second messenger systems. Tracer accumulation studies in trout (26) and kinetic analysis of Ca^{2+} pump activity in basolateral membranes from eel gills (29) have suggested that STC controls the permeability to Ca^{2+} of the apical membrane by controlling second messenger operated Ca^{2+} channels in the apical membrane (29). In eel gill cells STC reduced cAMP but had no significant effect on IP_3 (29). In trout gill cells, extracts of the corpuscles of Stannius reduced the levels of cyclic adenosine monophosphate (cAMP) (5). We here test the hypothesis that STC controls the apical membrane permeability for Ca^{2+} by looking for second messenger pathways for STC. We first considered cAMP as a second messenger for STC. We reasoned that the higher STC activity in seawater compared with that in freshwater counteracts the imminent threat of Ca^{2+} inflow (12) and could be mediated in two ways when STC acts through reduction of cAMP (5, 29). First, a long-term adaptation may involve decreased adenylate cyclase activity in gills from seawater-adapted fish. In trout (*Oncorhynchus mykiss*) basal adenylate cyclase activity in gill plasma membranes in seawater was half that in freshwater (10). Second, a short-term regulation through the immediate reduction of cellular cAMP by direct inhibition of adenylate cyclase and/or increased cAMP-phosphodiesterase activity was considered.

We performed our studies on tilapia (*Oreochromis mossambicus*), a well studied euryhaline teleost in which the mechanism of STC's action on second messengers is as yet unknown. We evaluated the effects of STC on the content of three second messengers in branchial epithelium, viz. cAMP, diacylglycerol (DAG), and inositol-1,4,5-trisphosphate (IP_3). Trout STC and/or eel STC_{1-20} (amino acid 1 to 20) were used to test the bioactivity of STC in tilapia gill cells or isolated plasma membranes because the glands from trout form a readily accessible source of STC.

These preparations were shown to reduce branchial Ca^{2+} uptake in tilapia (27), and thus possess the bioactivity of endogenous (tilapia) STC. This was confirmed by an experiment in which injection of antiserum to trout STC resulted in an increase Ca^{2+} influx in tilapia (3). The N-terminal 20 amino acids of STC are very well conserved between eel (2), trout (15) and tilapia (Verbost *et al.* in preparation). At the amino acid level there is 75% identity and 80% homology between eel STC_{1-20} and tilapia STC_{1-20}.

We included in our studies a previously defined low molecular weight protein from the corpuscles of Stannius, viz. teleocalcin (TC), that inhibited (low affinity) Ca^{2+}-dependent phosphatase activity and had no effect on whole-body Ca^{2+} influx (28). The purpose of this test was to find out whether products from the corpuscles of Stannius other than STC could also explain the reduction in cAMP found with extracts of the glands.

Materials and methods

Fish

Tilapia (*Oreochromis mossambicus*) from laboratory stock, 100 to 200 g in weight, were held in 100 l tanks, supplied with running tap water (0.7 mmol l⁻¹ Ca, 28 °C) under a photoperiod of 12 h of light alternating with 12 h of darkness. Animals were fed Trouvit fish pellets (Trouw & Co., Putten, The Netherlands), 1.5% body weight per day.

Analytical methods, hormones

Protein was determined with a commercial reagent kit (Bio-Rad, Mississauga, Ontario) according to the method of Bradford (1), using bovine serum albumin (BSA) (Bio-Rad) as a reference.

Radiotracer activities were determined with a Wallac 1410 liquid scintillation counter (Pharmacia LKB, Freiburg).

Trout STC (trSTC) was purified in two steps. The first step used concanavalin-A affinity chromatography as described in detail before (15). In the second step a size-separation was performed with the SMART system (Pharmacia LKB) on a Superdex 75 HR 10/30 column (high-performance gel filtration column). In all experiments we used trSTC which is bioactive in heterologous assays (16).

Peptide U, the 20 amino acid long N-terminal eel STC (eSTC) fragment, was synthesized, based on the cDNA amino acid sequence of the mature hormone, by Butkus and collaborators at the Howard Florey Institute peptide laboratory, Parkville, Australia (2). This peptide is a bioactive portion of STC (22, 27).

A partly purified low molecular weight fraction from trout Stannius corpuscles containing TC, a 3 kDa glycopeptide, was separated using a Sephadex G-25 column (fine 10×450 mm) (Pharmacia, Piscataway, NJ) with ammonium acetate as eluent (17, 28). The major peak, measured spectrophotometrically at 280 nm, was collected and freeze dried.

Isolation of basolateral membrane vesicles

Basolateral plasma membranes of tilapia gills were isolated as described by Verbost *et al.* (31). Protein recovery (around 2%) was not significantly different in freshwater (FW) preparations and seawater (SW) preparations. Isolated basolateral membranes were seven-to eightfold enriched in Na^+/K^+-ATPase activity. In SW preparations the purification of the membranes (expressed as the ratio of the Na^+/K^+-ATPase activity in the final pellet to that in the filament-homogenate) was slightly higher than in FW preparations (8.1 ± 1.5 µmol P_i h^{-1} mg^{-1} and 6.5 ± 1.6 µmol P_i h^{-1} mg^{-1}, respectively; not significantly different).

Isolation of gill cells

The gill cells were isolated following the recently described method (30) in which red blood cells are lysed and in which no enzymes are used to disrupt the tissue. The cells were washed in a 'fish Ringer' containing (in mmol l^{-1}) 114 NaCl, 4.2 KCl, 1.3 $CaCl_2$, 1.2 $MgSO_4$, 0.4 KH_2PO_4, 1.0 Na_2HPO_4, 0.1 $(NH_4)_2SO_4$, 13.1 $NaHCO_3$, 5.6 glucose, pH 7.4 (gassed with 95% O_2, 5% CO_2) and resuspended in the same medium for the assays. The red cell content of this preparation is less than 2%.

cAMP, DAG and IP_3 production in dispersed gill cells

cAMP Incubation was started by transferring cells (200 µl in fish Ringer, 500 µg protein) to tubes containing hormone (in 2 µl) and short vortex mixing. Cells were prewarmed for 2 min. The incubation temperature was 37 °C. Isobutyl methyl xanthine (IBMX) was not used. The reaction was stopped by mixing with 0.2 volume ice-cold 20% perchloric acid and the suspension was kept on ice for 20 min. The mixture was neutralized with 1.5 mol l^{-1} KOH containing 60 mmol l^{-1} Hepes buffer and phenol red as pH indicator. Proteins and precipitated $KCLO_4$ were sedimented by centrifugation at 9000 *g* for 2 min. Supernatants were used in the cAMP assay (see below).

DAG The same incubation procedure in fish Ringer was used as for cAMP measurements (see above). The incubation was stopped by transferring 400 µl suspension (around 1 mg protein) to an Eppendorf containing 800 µl ice-cold phosphate buffer; this mixture was centrifuged (9000 *g*, 1 min). Lipid was extracted from the cell suspension by addition of 0.5 ml methanol, 0.25 ml chloroform and 0.2 ml phosphate buffer (pH 7.4). This cocktail was vigorously vortexed and left to stand for 30 min at room temperature. After centrifugation (9000 *g*, 5 min) the supernatant was transferred to a tube containing 0.25 ml chloroform and 0.25 ml phosphate buffer. After centrifugation (5000 *g*, 5 min) the phases had been completely separated; the upper water phase was removed by aspiration with a tapered Pasteur's pipette, the lower chloroform phase was evaporated to dryness using a speed vacuum concentrator. The extracted lipids were kept under nitrogen until further assay (see below) according to the manufacturers prescription.

IP_3 For IP_3 the incubation procedure for the cAMP measurements (see above) was slightly modified: 100 µg instead of 500 µg protein was used, the medium contained 10 mM LiCl and the reaction was quenched in a different way. To ensure a quick stop the perchloric acid/cell-mixture was transferred to liquid nitrogen. After collecting all the samples this way, they were thawed, neutralized and centrifuged as above. From the resulting supernatant 100 µl was used in the IP$_3$ assay (see below).

Assays

Adenylate cyclase Adenylate cyclase (AC) activity in basolateral plasma membranes was assessed by measuring the formation of cAMP from ATP (cAMP was quantified with a binding assay, see below). The complete assay mixture (in 100 µl) contained: 10-20 µg protein, 60 mmol l⁻¹ sucrose, 18 mmol l⁻¹ NaCl, 3 mmol l⁻¹ MgCl$_2$, 20 mmol l⁻¹ Tris/HCl (pH 7.4), 0.8 mg ml⁻¹ BSA, 0.5 mmol l⁻¹ GTP, 1 mmol l⁻¹ ATP, 10 mmol l⁻¹ creatine phosphate, 0.3 mg ml⁻¹ creatine phosphokinase and 0.1 mmol l⁻¹ IBMX to block cAMP-dependent phosphodiesterase. Membranes and medium were mixed on ice and the reaction was started by transfer to 37 °C. The assay time was 20 min. For tilapia gill membranes the optimum temperature was 37 °C (at 28 and 22 °C cyclase activity was 22% and 39% lower, respectively, $n=5$). The reaction was stopped by the addition of 1 ml ice-cold 2-propanol (isopropyl alcohol); the tubes were vortexed (3 s) and placed on ice. Membranes were removed by centrifugation (5 min, 9000 *g*). The supernatant was transferred to clean tubes and dried in a speedvacuum. The dried samples were dissolved in 500 µl Tris-buffer (see assay of cAMP) from which 100 µl was used in the cAMP assay.

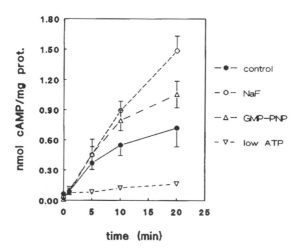

Fig. 1 AC activity in tilapia branchial basolateral membranes, measured as the ATP-dependent cAMP formation. NaF (5 mM) that stimulates G-proteins and the non-hydrolyzable GTP analogue GMP-PNP (0.5 mmol l-1) stimulate the cyclase activity. Lowering the ATP concentration 10 times (from 1.0 to 0.1 mmol l-1) reduced the activity by 90% (n=5, ±S.E.M).

The AC assay was validated as follows. A linear relation ($r=0.996$, $P<0.01$, four pairs of observations from three experiments) was found for 8-20 µg membrane protein and the amount of cAMP measured during a 20 min incubation. The cAMP production depended on the presence of GTP and ATP. The non-hydrolyzable GTP analog, guanylyl-imidodiphosphate (GMP-PNP) (Boehringer, Indianapolis, IN), significantly increased cAMP production (Fig. 1). The optimum ATP concentration was 1.0 mmol l^{-1}, indicating a millimolar affinity for ATP, also found for trout gill membranes (0.5 mmol l^{-1} (11)). A twofold stimulation of cAMP production was observed by addition of 5 mmol l^{-1} NaF to stimulate the G-protein coupled AC. Furthermore, no cAMP was found in membrane preparations that had been boiled for 10 min and this shows that the endogenous cAMP in this membrane preparation is negligible. Also, the STC preparations were devoid of cAMP and did not interfere with the cAMP binding assay: when STC was added after the incubation had been stopped, no effect was seen on the apparent cAMP production (25 µg ml^{-1} TC or 100 µg ml^{-1} STC added afterwards resulted in $107\pm16\%$ and $100\pm13\%$ ($n=4$) of the control value, respectively).

cAMP-phosphodiesterase (PDE) The enzyme activity was measured by a modification of the procedure of Teo *et al.* (24). In this assay the membrane mediated conversion of cAMP into 5'AMP is determined. To this end the P_i release, when 5'AMP is broken down into adenosine and P_i by 5'nucleotidase (N-5880, Sigma, Chem. Co., Deisenhofen), is measured. The assay mixture (in 250 µl) contained: 10-20 µg membrane protein, 40 mmol l^{-1} imidazole, 20 mmol l^{-1} Mg-acetate, 40 mmol l^{-1} Tris/HCl (pH 7.5), 2 units ml^{-1} 5'nucleotidase, 0 or 2.5 mmol l^{-1} cAMP. The incubation was started by addition of the membranes to the medium (vortex) and transfer of the mixture to 37 °C. The assay time was 30 min. The reaction was stopped by the addition of 0.5 ml ice trichloric acid (8.6%); the P_i concentration was determined colorimetrically according to Fiske and SubbaRow (4).

The assay was validated as follows. One prerequisite for this assay is that the test substance does not affect the 5'nucleotidase activity. The test-substances did not affect the dephosphorylation rate of 5'AMP by 5'nucleotidase. Furthermore, a differential assay was performed with or without cAMP and the differences in P_i liberated was taken as a measure for the phospodiesterase (PDE) activity. Purified PDE from Sigma Cem. Co. (P-0134) was used as a standard.

cAMP Cyclic AMP was quantified using a cAMP binding protein kit (TRK.432, Amersham, Bucks). The detection limit of the assay is 1 pmol per tube. Data were expressed in pmol cyclic nucleotide per mg protein present in the propanol extract of the tissue.

Diacylglycerol The branchial content of sn-1,2-DAG was determined using a commercial reagent kit (RPN 2009, including Amprep columns, Amersham). Extracted lipids were resuspended in the detergent mix provided with the DAG reagent kit and serially diluted therein. The quantity of membranes used in the assay was equivalent to 0.5-1.0 mg BSA equivalents membrane protein.

The radioenzymatic assay employs DAG kinase from *E. coli* to convert DAG, solubilized in a defined mixed micelle condition, into [^{32}2P]-phosphatidic acid (PA) in the presence of [^{32}P]- ATP (PB.108, Amersham). The reaction product [^{32}P]-PA is extracted by Amprep column chromatography and the amount of [^{32}P]-PA determined by liquid scintillation counting. The amount of DAG in the sample is calculated from the [^{32}P]-PA and the specific activity of the [^{32}P]- ATP used in the assay. The sensitivity of the assay allows the detection of 10 pmol per tube. Data are presented in nmol DAG per mg protein, referring to the amount of membranes (in mg protein equivalents) used for lipid extraction.

IP₃ The Amersham IP$_3$ assay kit, TRK.1000, applying the IP$_3$ binding protein method was used following the extraction procedure as described above. Data were expressed in pmol cyclic nucleotide per mg protein present in the perchloric acid extract of the tissue.

Immunostaining of gills

Gill samples were obtained from the third gill arch on the left side. Tissues were fixed for 24 h in Bouin's fixative, washed three times in 70% ethanol and kept in 90% ethanol until further processing. After dehydration in serial alcohols, the samples were embedded in paraffine. Longitudinal 5 μm sections were cut (Reichert Jung Biocut Model 1330, Germany). Following deparaffination sections were treated with 1% hydrogen peroxide in methanol (30 min, room temperature) to neutralize endogenous peroxidase activity. The first antibody (Chemicon, Temecula, CA, USA) was applied in 1:1000 dilution, and for the staining an ABC-kit (Vectastain, Burlingame, CA, USA) with 3,3'-diaminobenzidine as peroxidase substrate was used.

Statistics and calculations

Values are presented as means±standard deviation, unless otherwise stated. The unpaired or paired Student's *t*-test was applied where appropriate to assess statistical significance of differences between means. A P-value<0.05 was taken to indicate significance.

Results

Adenylate cyclase in gills and intestine of FW and SW tilapia

Basal and NaF stimulated AC activity in basolateral membranes from FW gills were significantly higher than in membranes from SW gills (Fig. 2).

Effects of STC on cellular cAMP

In cells from FW gills, trSTC had a biphasic effect; it reduced cAMP formation atpicomolar concentrations but the inhibitory effect reduced with increasing hormone concentrations (in the picomolar range) and was absent at nanomolar concentrations (Fig. 3).

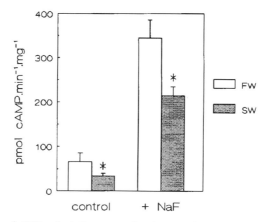

Fig. 2 Effect of SW adaptation on adenylate cyclase activity in basolateral membranes from gills and intestine of tilapia (n=5-8, ±S.E.M). $*$: significantly lower than in FW membranes.

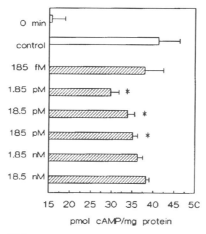

Fig. 3 Effects of trSTC on gill cell cAMP content. Cells were incubated for 10 min at 28 $\times$C (n=5, ±S.E.M). $*$: significantly lower than control.

Immunostaining of cAMP in gills and intestine from FW and SW tilapia
Immunostaining of the gills with anti-cAMP labels cells that, based on their location and morphology, appear to be chloride cells (Fig. 4). Applying this staining simultaneously to FW and SW gills as a semi-quantitative method, it follows that the relative concentration of cAMP in the gill cells was higher in FW than in SW gills. The number of chloride cells in FW and SW gills is similar in these fish, but the cells are significantly bigger in SW than in FW gills (confocal laser scanning microscopy results, personal observations).

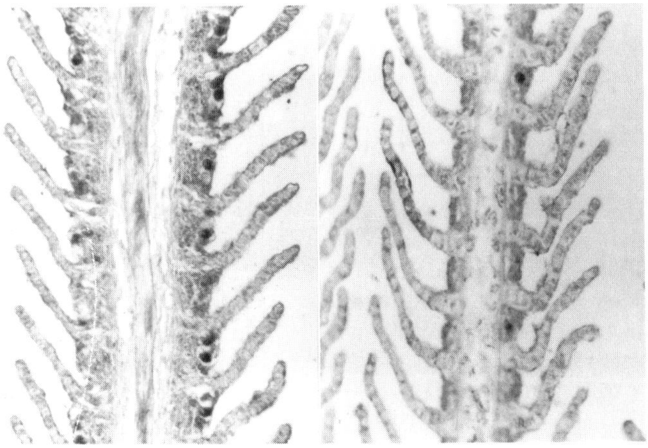

Tilapia

FW   SW

anti cAMP 1:900 (Magn. 250x)

Fig. 4 Anti-cAMP stained gill sections of tilapia adapted to FW or SW.

Effects of trSTC, eSTC$_{1-20}$, and trTC on AC

Purified trSTC had no effect on NaF-stimulated plasma membrane AC in basolateral membranes from FW tilapia gills tested at 18.5 pmol l^{-1} to 185 nmol l^{-1}, increased the activity by 35% at 370 nmol l^{-1}, and had no significant effect at 0.74 to 1.48 μmol l^{-1} (Fig. 5). In the absence of NaF, trSTC tested in a range from 1.85 pmol l^{-1} to 185 nmol l^{-1} (which equals 0.10 ng ml^{-1} to 10 μg ml^{-1}) had no effect (Fig. 5). The peptide eSTC$_{1-20}$, tested in a range from 1.85 nmol l^{-1} to 100 μmol l^{-1}, had no effect on AC (Fig. 5); the 18% increase with 10 μmol l^{-1} of the peptide was not significant.

Our trTC preparation inhibited AC activity (Fig. 6): at 40 μg ml^{-1} trTC AC was inhibited 23±4% at non-stimulated conditions and 14±3% under NaF-stimulated conditions. Extracts of corpuscles of Stannius also inhibited the AC activity: 13% at 25 μg ml^{-1} and 21% inhibition at 50 μg ml^{-1} (in the presence of NaF).Thus, the AC inhibitory action of the extract can be attributed to the activity of TC.

Effect of trSTC on cAMP-phosphodiesterase

The PDE inhibitor IBMX (at 1 mmol l^{-1}) significantly reduced the cAMP-phosphodiesterase activity (cAMP-PDE) in basolateral membranes from FW tilapia gills by 42%. trSTC at 1.85 pmol l^{-1} to 185 nmol l^{-1} did not influence cAMP-PDE (Fig. 7).

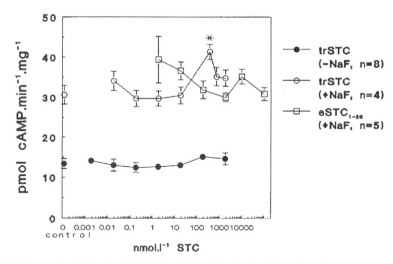

Fig. 5 Effects of trSTC and eSTC1-20 on AC activity in basolateral membranes from the gills, determined in the presence or absence of NaF (n values in figure legend). ✻: significantly lower than control (indicated at Y-axis).

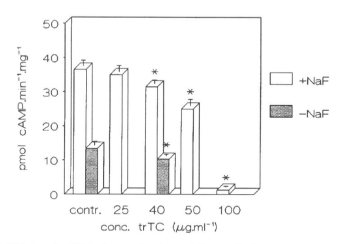

Fig. 6 Effects of trTC (a low molecular weight protein from trout corpuscles of Stannius) on AC activity in basolateral membranes from the gills, measured in the presence or absence of NaF (n=4-6, ±S.E.M). ✻: significantly lower than respective control.

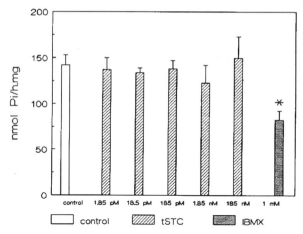

Fig. 7 Effects of trSTC and IBMX on the cAMP-phosphodiesterase activity in basolateral membranes from the gills (n=4, ±S.E.M). ✱: significantly lower than control.

Effect of trSTC and on cellular DAG and IP$_3$.

Table 1 Effects of trSTC on diacylglycerol-, IP3- and cAMP-production in isolated gill cells (n=7, ±S.E.M).

	DAG 10 min inc. (%)	IP$_3$ 0.5 min inc. (in pmol mg^{-1})	cAMP 10 min inc. (in pmol mg^{-1})
control	100±6.1	53.3±4.5	3.53±0.42
trSTC 18.5 nmol l^{-1}	133±15.8*	50.6±4.5	3.62±0.59
eSTC$_{1-20}$ 185 nmol l^{-1}	146±16.9*	51.2±4.6	3.09±0.40

*significantly different from control; *P<0.05.*

trSTC (18.5 nmol l^{-1}) and the bioactive N-terminal fragment eSTC$_{1-20}$ (185 nnmol l^{-1}) increased DAG production in dispersed tilapia gill cells during a 10 min incubation at 28 °C (Table 1). The absolute DAG levels reached in the presence of eSTC$_{1-20}$ were significantly higher than in controls (7.54±1.33 nmol DAG mg^{-1} protein versus 5.31±0.73 nmol DAG mg^{-1} protein). Absolute DAG levels in the presence of trSTC were not significantly different from controls due to considerable variation between the different preparations. The absolute numbers varied from 5 to 10 nmol mg^{-1} protein.

In the same preparations of gill cells this trSTC concentration did not effect cAMP production over a 10 min period nor IP_3 production measured over a 30 s period (Table 1).

Discussion

The results show that in gills from tilapia adapted to SW the AC activity is lower and cellular cAMP in the chloride cells is lower than in FW tilapia. AC activity is measured in a membrane fraction enriched in Na^+/K^+-ATPase and will therefore mainly derive from the chloride cells (7, 8).

Purified trout-STC (trSTC) reduced production of cAMP in gill cells but only at pico-molar concentrations. At a relatively high concentration (370 nmol l^{-1}), trSTC stimulated adenylate cyclase activity. ELISA and RIA measurements have suggested, however, that nano-molar concentrations (1-5 nmol l^{-1}) are present in the blood (12, 21). It is difficult to predict what action will prevail because the actual concentrations locally at the gill cell membranes are unknown. When a crude extract of trout corpuscles is used in the same assay, 2 µg ml^{-1} extract, containing approximately 500 ng ml^{-1} STC, results in $47\pm13\%$ ($n=5$) reduction in cAMP formation. trTC at 6 µg ml^{-1} lowers cAMP production by $69\pm10\%$ ($n=5$) (data not presented in results). It is concluded from these results that the reduction in cellular cAMP induced by extracts is mainly explained by the effect of TC.

Tested over a wide concentration range, trSTC as well as the N-terminal fragment of $eSTC_{1-20}$ had no inhibitory effect on the AC activity in the membranes of the gill cells. However, trTC did inhibit the AC. Obviously, the inhibition of the cyclase by TC can explain the reduction in cellular cAMP found with a homogenate of Stannius corpuscles (5); this study). Although STC did not reduce AC, picomolar STC was able to reduce cAMP formation and it was tested whether this could be caused by an activation of the cAMP-PDE. But no effects of STC on the PDE activity in the basolateral membranes were found. In conclusion, the mechanism for the reduced cAMP formation in reaction to STC is unknown at present. The reduction in cAMP cannot be explained by an increase in release of cAMP either since the cell samples were not separated from the medium before cAMP extraction. The results indicate the existence of an indirect effect.

The reduction of the AC activity by TC was stronger in the non-stimulated conditions. This is comprehensible, assuming that the inhibition occurs via an inhibitory GTP-bindingprotein (G_i) and when considering the mechanism of action for NaF, through stimulation of GTP-binding proteins (G_s). The inhibition could be partly masked in the presence of NaF because both G_i and G_s α-units use the same β and γ units (23). STC was also tested in the absence of NaF thus excluding this possible masking effect, but still no inhibitory effect was observed in a concentration range from 2 pmol l^{-1} to 2 µmol l^{-1} (results not shown).

There was a small but consistent increase in AC activity at STC concentrations of around 0.4 µmol l^{-1}. At higher concentrations (≥ 0.8 µmol l^{-1}) the stimulatory effect was non-significant and at lower concentrations (≤ 20 nmol l^{-1}) there was no effect at all. Plasma STC levels have been reported for several fish species to be between 1 and

5 nmol l^{-1}, as determined with the ELISA (21). It is hard to tell, therefore, whether the stimulatory effect is of physiological importance. However, there is some recent work, albeit on another species, in favour of such a cAMP signal in the inhibition of Ca^{2+} uptake. When tested in an Ussing chamber setup, dibutyryl cAMP reduces transcellular Ca^{2+} uptake in killifish opercular membranes (20, 32).

Taken together, all the results obtained with purified STC lead to the conclusion that there is no solid evidence to propose a reduction in cAMP as a second messenger signal for STC. It could very well be that other factors (like TC) from the corpuscles of Stannius are able to reduce cellular cAMP, but the possible function of such an action is unknown. It is also significant that trTC has no effect on Ca^{2+} influx in tilapia (28). At present it seems more likely that STC leads to an increase in cAMP which in turn reduces Ca^{2+} influx. Furthermore, a second messenger signal does not often come alone, and we looked for other second messengers as well.

At the physiological concentration of 18.5 nmol l^{-1} trSTC increased DAG levels in the tilapia gill cells *in vitro*. Congruent with the other results this concentration had no effect on cAMP. Further, we found no effect on inositol 1,4,5-triphosphate (IP$_3$) levels. Normally, a stimulation of DAG production coincides with an IP$_3$ response. There are two possible explanations for the absence of an IP$_3$ effect. One may relate to the difficulty in measuring IP$_3$. Obviously, the breakdown of IP$_3$ could be a problem and it is very likely that only the chloride cells react to STC, which make up only about 10% of the total cell population, thus making it increasingly difficult to measure a small peak. The other possibility is that DAG is being produced through a route other than that described above (13). Conversely, a Ca^{2+} signal could be involved as indicated by Ussing chamber experiments with killifish operculum membrane. Thapsigargin, a blocker of endoplasmic reticulum Ca^{2+} pumps (25) that can be used to raise intracellular Ca^{2+} levels, reduced transcellular influx of Ca^{2+} in opercular membranes (20, 32).

Perhaps we have to adopt a view in which cAMP is not so much a second messenger in the regulation of Ca^{2+} uptake but more a second messenger for Na$^+$ and Cl$^-$ uptake (in FW) and secretion (in SW). In FW-adapted fish osmoregulation is believed to be under neurohypophyseal control (18). Receptors for the neurohypophyseal peptides arginine vasotocin and isotocin are coupled to the AC via G$_i$, the inhibitory transducer protein (11). In contrast to the situation in FW, the gills in SW-adapted fish secrete NaCl. Increased levels of intracellular cAMP stimulate chloride secretion (6) whereas increased Ca^{2+} levels inhibit chloride secretion (19). These reports suggest that the NaCl exchange is intracellularly controlled by cAMP and possibly the basal levels in FW- and SW-adapted fish are dictated by the necessity to maintain control over the transport of these monovalent ions. However, if interactions between Na$^+$ and Ca^{2+} transport exist the picturewill become much more complex. Indeed, there are reasons to suggest such an interplay. Augmentation of intracellular cAMP in opercular epithelia from SW-adapted killifish stimulates Cl$^-$ secretion by increasing the rate of Na$^+$/K$^+$/2Cl$^-$ cotransport at the basal side of the chloride cell which in turn will reduce the Na$^+$ gradient across this membrane. This would reduce the driving force of Na^+/Ca^{2+} exchange at the basolateral membrane. In SW and FW opercular membranes from killifish, Ca^{2+} influx was 85-90% dependent on Na^+/Ca^{2+} exchange in the basolateral membrane (32), the presence of which was also shown biochemically in tilapia and killifish gills (31, 32). Also in FW killifish opercular membrane augmentation of cAMP results in a SW-like short-circuit current and, as in SW membranes, was invariably associated with a decrease in Ca^{2+} influx (19). So indeed, there is interplay, at least at the second messenger level, between transcellular transport of monovalent ions and Ca^{2+} and more evidence for such interactions can now be obtained using the opercular epithelium.

Acknowledgements

The work was carried out in Nijmegen and supported by a fellowship of the Royal Netherlands Academy of Arts and Sciences (KNAW) to P M V; G F received a travel grant from the KNAW. The research was also supported by an operating grant (A6246) from the Natural Sciences and Engineering Council of Canada to J C F. The authors thank Mr F A T Spanings for excellent organization of fish husbandry.

References

1   Bradford MM (1976) A rapid and sensitive method for the quantitation of microgram quantities of protein utilizing the principle of protein-dye binding. *Analytical Biochemistry* **72** 248-254.

2   Butkus A, Roche PJ, Fernley RT, Haralambidis J, Penschow JD, Ryan GB, Trahair JF, Tregear GW and Coghlan JP (1987) Purification and cloning of a corpuscle of Stannius protein from *Anguilla australis. Molecular and Cellular Endocrinolgy* **54** 123-133.

3   Fenwick JC, Flik G and Verbost PM (1995) A passive immunization technique against the teleost hypocalcemic hormone stanniocalcin provides evidence for the cholinergic control of stanniocalcin release and the conserved nature of the molecule. *General and Comparative Endocrinology* **98** 202-210.

4   Fiske CH and SubbaRow Y (1925). The colorimetric determination of phosphorus *Journal of Biological Chemistry* **66** 375.

5   Flik G (1990) Hypocalcin physiology. *Progress in Comparative Endocrinology*, pp 578-585. New York: Wiley-Liss.

6   Foskett JK, Bern HA, Machen TE and Conner M (1983) Chloride cells and the hormonal control of teleost fish osmoregulation. *Journal of Experimental Biology* **106** 255-281.

7   Flik G, van Rijs JH and Wendelaar Bonga SE (1985) Evidence for a high-affinity Ca^{2+}-ATPase activity and ATP-driven Ca^{2+}-transport in membrane preparations of the gill epithelium of the cichlid fish *Oreochromis mossambicus. Journal of Experimental Biology* **119** 335-347.

8   Flik G, Wendelaar Bonga SE and Fenwick JC (1985) Active Ca^{2+} transport in plasma membranes of branchial epithelium of the North-American eel *Anguilla rostrata* LeSueur. *Biology of the Cell* **55** 265-272.

9   Flik G, Labedz T, Lafeber FPJG, Wendelaar Bonga SE and Pang PKT (1989) Studies on teleost corpuscles of Stannius: physiological and biochemical aspects of synthesis and release of hypocalcin in trout, goldfish and eel. *Fish Physiology and Biochemistry* **7** 343-349.

10 Guibbolini ME and Lahlou B (1987) Adenylate cyclase activity in fish gills in relation to salt adaptation. *Life Sciences* **41** 71-78.

11 Guibbolini ME and Lahlou B (1987) Neurohypophyseal peptide inhibition of adenylate cyclase activity in fish gills The effects of environmental salinity. *FEBS Letters* **220** 98-102.

12 Hanssen RGJM, Mayer-Gostan N, Flik G and Wendelaar Bonga SE (1993) Stanniocalcin kinetics in freshwater and seawater european eel (*Anguilla anguilla*). *Fish Physiology and Biochemistry* **10** 491-496.

13 Hata Y, Ogata E and Kojima I (1989) Platelet-derived growth factor stimulates synthesis of 1,2-diacylglycerol from monoacylglycerol in Balb/c 3T3 cells. *Biochemical Journal* **262** 947-952.

14 Lafeber FPJG, Flik G, Wendelaar Bonga SE and Perry SF (1988) Hypocalcin from Stannius corpuscles inhibits gill calcium uptake in trout. *American Journal of Physiology* **254** R891-R896.

15 Lafeber FPJG, Hanssen RGJM, Choy YM, Flik G, Hermann-Erlee MPM, Pang PKT and Wendelaar Bonga SE (1988) Identification of hypocalcin (teleocalcin) isolated from trout Stannius corpuscles. *General and Comparative Endocrinology* **69** 19-30.

16 Lafeber FPJG, Hanssen RGJM and Wendelaar Bonga SE (1988) Hypocalcemic activity of trout hypocalcin and bovine parathyroid hormone in stanniectomized eels. *Journal of Experimental Biology* **140** 199-208.

17 Ma SWY and Copp DH (1978) Purification, properties and action of a glycopeptide from the corpuscles of Stannius which affects calcium metabolism in the teleost. *Comparative Endocrinology,* pp 283-286. Eds PJ Gaillard and HH Boer. Amsterdam: Elsevier.

18 Maetz J and Lahlou B (1974) Actions of neurohypophyseal hormones in fishes. *Handbook of Physiology: Endocrinology, the pituitary gland and its neuroendocrine control,* vol IV, pp 521-544. Washington DC: American Physiological Society section 7.

19 Marshall WS, Bryson SE and Garg D (1993) Alpha 2-adrenergic inhibition of Cl-transport by opercular epithelium is mediated by intracellular Ca2+. *Proceedings of the National Academy of Sciences of the USA* **90** 5504-5508.

20 Marshall WS, Bryson SE, Burghardt JS and Verbost PM (1995) Ca^{2+} transport by ionocytes in opercular epithelium of the euryhaline teleost, *Fundulus heteroclitus. Journal of Comparative Physiology B: Biochemical Systemic and Environmental Physiology* **165** 268-277.

21 Mayer-Gostan N, Flik G and Pang PKT (1992) An enzyme-linked immunosorbent assay for stanniocalcin, a major hypocalcemic hormone in teleosts. *General and Comparative Endocrinology* **86** 10-19.

22 Milliken CE, Fargher RC, Butkus A, McDonald M and Copp DH (1990) Effects of synthetic peptide fragments of teleocalcin (hypocalcin) on calcium uptake in juvenile rainbow trout (*Salmo gairdneri*). *General and Comparative Endocrinology* **77** 416-422.

23 Neer EJ and Clapham DE (1988) Roles of G protein subunits in transmembrane signalling. *Nature* **333** 129-134.

24 Teo TS, Wang TH and Wang JH (1973) Purification and properties of the protein activator of bovine heart cyclic adenosine 3',5'-monophosphate phosphodiesterase. *Journal of Biological Chemistry* **248** 588-595.

25   Thastrup O, Cullen PJ, Drobak BK, Hanley MR and Dawson AP (1990) Thapsigargin, a tumor promotor, discharges intracellular Ca^{2+} stores by specific inhibitionof the endoplasmic reticulum Ca^{2+}-ATPase. *Proceedings of the National Academy of Sciences of the USA* **87** 2466-2470.

26   Verbost PM, Van Rooij J, Flik G, Lock RAC and Wendelaar Bonga SE (1989) The movement of cadmium through freshwater trout branchial epithelium and its interference with calcium transport. *Journal of Experimental Biology* **145** 185-197.

27   Verbost PM, Butkus A, Atsma W, Willems P, Flik G and Wendelaar Bonga SE (1993) Studies on stanniocalcin: characterization of bioactive and antigenic domains of the hormone. *Molecular and Cellular Endocrinolgy* **93** 11-16.

28   Verbost PM, Butkus A, Willems P and Wendelaar Bonga SE (1993) Indications for two bioactive principles in the corpuscles of Stannius. *Journal of Experimental Biology* **177** 243-252.

29   Verbost PM, Flik G, Fenwick JC, Greco A-M, Pang PKT and Wendelaar Bonga SE (1993) Branchial calcium uptake: possible mechanisms of control by stanniocalcin. *Fish Physiology and Biochemistry* **11** 205-215.

30   Verbost PM, Flik G and Cook H (1994) Isolation of gill cells. *Biochemistry and Molecular Biology of Fishes,* vol 3, pp 239-247. Eds PW Hochachka and TP Mommsen. Amsterdam: Elsevier.

31   Verbost PM, Schoenmakers TJM, Flik G and Wendelaar Bonga SE (1994) Kinetics of ATP- and Na^+-gradient driven Ca^{2+} transport in basolateral membranes from gills of freshwater- and seawater-adapted tilapia. *Journal of Experimental Biology* **186** 95-108.

32   Verbost PM, Bryson SE, Wendelaar Bonga SE and Marshall WS (1995) Na^+ dependent Ca^{2+} uptake in isolated opercular epithelium of *Fundulus heroclitus. American Journal of Physiology* (In Press).

33   Wagner GF, Hampong M, Park CM and Copp DH (1986) Purification,characterization, and bioassay of teleocalcin, a glycoprotein from salmon corpuscles of Stannius. *General and Comparative Endocrinology* **63** 481-491.

34   Wendelaar Bonga SE and Pang PKT (1986) Stannius corpuscles. *Vertebrate Endocrinology, Fundamentals and Biomedical Implications*, pp 439-464. Eds PKT Pang and MP Schreibman. New York: Academic Press.

35   Wendelaar Bonga SE and Pang PKT (1989) Pituitary hormones. *Vertebrate Endocrinology: Fundamentals and Biomedical Implications Vol 3: Regulation of Calcium and Phosphate.* Eds PKT Pang and MP Schreibman. San Diego: Academic Press.

The Comparative Endocrinology of Calcium Regulation
Eds C Dacke, J Danks, I Caple & G Flik, pp 71-74
Journal of Endocrinology Ltd, Bristol (1996)

Evolutionary aspects of vitamin D metabolism in vertebrates, with an emphasis on fish and fetal rats

T Kobayashi, A Takeuchi, T Okano, H Sekimoto and Y Ishida

Department of Hygienic Sciences, Kobe Pharmaceutical University, Kobe
658, Japan

Introduction

It is well documented that vitamin D_3 (animal form of vitamin D) is metabolized to 25-hydroxyvitamin D_3 (25-D_3) in the liver and subsequently to 1α,25-dihydroxyvitamin D_3 (1,25-D_3) or 24R,25-dihydroxyvitamin D_3 (24,25-D_3) in the kidney of mammals (1). Since 1,25-D_3 has the most potent activity among the metabolites, it is considered the active form of vitamin D_3; there is consensus that in mammals the kidney is the site of the 25-D_3-1α-hydroxylase (25-D_3-1α-OHase). In contrast, we have found that in fish the liver contains 25-D_3-1α-OHase next to the kidney, and this situation is different from mammals where the enzyme is restricted to the kidney (4). We have shown that the activity of the enzyme in fish liver is higher than in the kidney. These findings suggest that the mechanism of vitamin D_3 metabolism may have changed during the evolution of vertebrates. Accordingly, we wondered whether the enzyme exists in the liver of fetal rats as it does in fish. From these viewpoints, we studied vitamin D_3 metabolism in a phylogenetic context.

Concentrations of vitamin D_3 and its metabolites in fish plasma

Concentrations of vitamin D_3 and its metabolites in fish plasma were assayed as previously reported (3). As shown in Table 1, the concentrations of vitamin D_3 and 1,25-D_3 in fish plasma were higher than the respective values in human plasma, but those of 25-D_3 and 24,25-D_3 in almost all the fish plasma (except yellowtail) were under the limit of detection. The results are quite different from those observed in the plasma of mammals, showing that the 25-D_3 concentration is highest among the metabolites. Why do fish contain higher concentrations of 1,25-D_3 in plasma without 25-D_3? This is an important question that may be solved through a comparative biochemical approach and in order to solve it we have performed the following studies.

Existence of 25-D_3-1α-OHase in fish liver

When a liver homogenate of vitamin D-deficient carp was incubated with [^{3}H]-25-D_3 and subsequently applied to HPLC, a peak corresponding to 1,25-D_3 was recognized in the chromatogram. The peak comigrated with pure 1,25-D_3. The peak was not observed when the homogenate was previously heated at 100 °C for 15 min,

Table 1 Vitamin D_3 and its metabolites in fish plasma

Sample	D_3 (ng/ml)	$25\text{-}D_3$ (ng/ml)	$24,25\text{-}D_3$ (ng/ml)	$1,25\text{-}D_3$ (pg/ml)
Carp	35	ND	ND	174
Gibel	22	ND	ND	46
Japanese eel	29	ND	ND	109
Red sea bream	5	ND	ND	190
Yellowtail	202	2	1	54
Striped jack	34	ND	ND	84
Chub mackerel	12	ND	ND	85
File fish	23	ND	ND	202
Greenling	25	ND	ND	256
Bastard halibut	127	ND	ND	192
Human	2	21	1	42

Note: ND = not detected.

suggesting that an enzymatic reaction is involved in its production. These results suggest that carp liver contains $25\text{-}D_3\text{-}1\alpha\text{-}OHase$, which is a new finding. The formation of $1,25\text{-}D_3$ was reconfirmed by the retention times on HPLC using two kinds of column, thermal isomerization of $1,25\text{-}D_3$ into its pre-isomer (2), gas chromatography/mass spectrometry (GC-MS) and the reaction with the vitamin D receptor (VDR). All results confirmed that the incubation mixture of $25\text{-}D_3$ with the carp liver homogenate contained $1,25\text{-}D_3$ and the existence of $25\text{-}D_3\text{-}1\alpha\text{-}OHase$ in the carp liver seems undeniable. Investigations into the subcellular location in the liver have shown that $25\text{-}D_3\text{-}1\alpha\text{-}OHase$ exists and mainly resides in the mitochondrial fraction, while vitamin $D_3\text{-}25\text{-}OHase$ mainly resides in the microsomal fraction. When similar experiments were carried out with the kidney of vitamin D-deficient carp, the existence of $25\text{-}D_3\text{-}1\alpha\text{-}OHase$ was confirmed in the mitochondrial fraction, while vitamin $D_3\text{-}25\text{-}OHase$ was not found in the kidney. The enzymatic activity in the liver was stronger than that in the kidney.

Comparative studies on vitamin D metabolism

The experiments mentioned for liver and kidney of carp were then carried out with tissues of tadpoles and bullfrogs (amphibians), soft-shelled turtles (reptile) and rats (mammals), and the enzymatic activities were compared to assess a phylogenetic series of vitamin D metabolism (Fig. 1). The enzymatic formation of 1,25-D_3 was observed in the liver of carp, tadpoles and bullfrogs but not in that of the soft-shelled turtle or the rat, while 1,25-D_3 was formed in the kidney of all species. The activity of 25-D_3-1α-OHase in the liver gradually decreased during vertebrate evolution and it had completely disappeared in the soft-shelled turtle and rat; in the kidney it increased inversely to the activity in the liver. Interestingly, the vitamin D-binding protein (DBP) carrying 25-D_3 could not be detected in the plasma of fish and amphibians, but was found in the reptiles and mammals. These results suggest that 25-D_3-1α-OHase activity has moved from liver to kidney during the evolution of vertebrates and that this change coincides with the appearance of plasma DBP, which specially transports 25-D_3 .

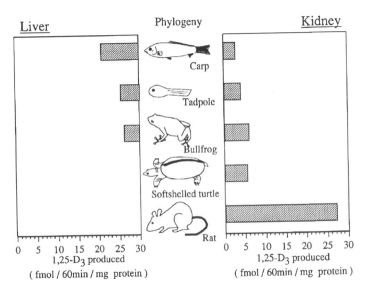

Fig. 1 Comparison of 25-D_3-1α-hydroxylase activity in liver and kidney of various vertebrates.

Enzymatic formation of 1,25-D_3 from 25-D_3 in the liver of fetal rats

Since ontogeny sometimes follows phylogeny as indicated above, we then studied the manner of vitamin D metabolism in fetal, neonatal and mature rats (5). Vitamin D-deficient fetal rats were obtained from pregnant mothers. Liver homogenate of fetal rats was incubated with [^{3}H]-25-D_3 and subjected to the procedure described above. The HPLC profiles show that the fetal rat liver homogenate indeed produces a peak corresponding to pure 1,25-D_3. Since the peak corresponding to 1,25-D_3 was not observed when the homogenate was pre-heated, it was concluded that an enzymatic reaction underlies the production of this peak. The formation of 1,25-D_3 from 25-D_3 in the homogenate of fetal rat liver was further confirmed by HPLC, thermal isomerization into its pre-isomer, binding affinity to VDR and GC-MS as described above for fish.

Changes in the enzymatic activity of 25-D$_3$-1α-OHase in the liver and kidney of fetal, neonatal and mature rats were then examined. As shown in Fig. 2, the enzymatic activity in the fetal rat (21 days) liver was almost the same as that in the kidney, but the enzyme was absent from the liver of the mature rat and predominantly existed in the kidney. Therefore, the activity in the liver may disappear while it increases in the kidney with growth towards maturity. The results obtained in this comparative study on 25-D$_3$-1α-OHase activity support the thesis that ontogeny is a recapitulation of phylogeny.

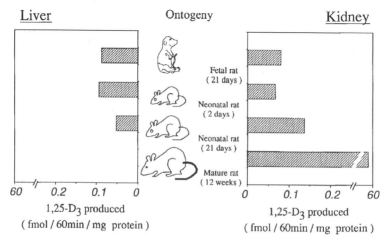

Fig. 2 Comparison of 25-D$_3$-1α-hydoxylase activity in liver and kidney of fetal, neonatal and mature rats.

References

1   Holick MF, Schnoes HK, DeLuca HF, Suda T & Cousins RJ (1971) Isolation and identification of 1,25-dihydroxycholecalciferol A metabolite of vitamin D active in intestine. *Biochemistry* **10** 2799-2804.

2   Okano T, Tsugawa N & Kobayashi T (1989) Identification and determination of 1α,25-dihydroxyvitamin D$_3$ in rat skin by high-performance liquid chromatography. *Journal of Chromatography* **493** 63-70.

3   Takeuchi A, Okano T, Tsugawa N, Tasaka Y, Kobayashi T, Kodama S & Matsuo T (1989) Effects of ergocalciferol supplementation of vitamin D and its metabolites in human milk. *Journal of Nutrition* **119** 1639-1646.

4   Takeuchi A, Okano T & Kobayashi T (1991) The existence of 25-hydroxyvitamin D$_3$-1α-hydroxylase in the liver of carp and bastard halibut. *Life Sciences* **32** 275-282.

5   Takeuchi A, Okano T, Sekimoto H & Kobayashi T (1994) The enzymatic formation of 1α,25-dihydroxyvitamin D$_3$ from 25-hydroxyvitamin D$_3$ in the liver of fetal rats. *Comparative Biochemistry and Physiology* **109C** 1-7.

The Comparative Endocrinology of Calcium Regulation
Eds C Dacke, J Danks, I Caple & G Flik, pp 75-84
Journal of Endocrinology Ltd, Bristol (1996)

Calcium regulation by the vitamin D_3 system in teleosts, with special emphasis on the intestine

K Sundell, D Larsson and B Th Björnsson

Fish Endocrinology Laboratory, Department of Zoophysiology, Göteborg University, Medicinaregatan 18, S-413 90 Göteborg, Sweden

Introduction

Cod liver oil has traditionally been utilized in Northern countries to prevent the bone-debilitating disease rickets. The active substance in cod liver oil and other fats (43), as well as in some other food substances after ultraviolet (UV) irradiation (30, 57), was found to be a new fat-soluble vitamin. It was the fourth vitamin discovered and thus given the letter D (42). It is now known that the vitamin D ingested in the diet or obtained from the skin after UV irradiation, is not the form active in calcium homeostasis. The existence of a more polar metabolite responsible for the biological actions was first postulated in 1968 by Haussler *et al.* (24). This metabolite is formed by two successive hydroxylations of vitamin D_3, producing 1,25-dihydroxyvitamin D_3 $(1,25(OH)_2D_3)$. This metabolite is now considered to be hormonally active in calcium regulation of mammals and birds (29).

For the field of comparative endocrinology, one obvious question arises: apart from being beneficial to humans, what is the physiological role of the vitamin D_3 pool in the fish itself? This chapter will review the physiological role of the vitamin D_3 system in calcium regulation of teleosts, with special emphasis on the intestine.

Origin of vitamin D_3 in teleosts

Since Bills first raised the issue in 1927 (8), the origin of the high vitamin D_3 content of fish liver has remained unclarified. However, three possible ways for fish to obtain their vitamin D_3 have been suggested.

One way in which the vitamin can be obtained is through ingestion of food. The original food sources for fish, phytoplankton and zooplankton, have been investigated for their vitamin D content since the 1920s. Early studies revealed that most phytoplankton contained little or no antirachitic activity (16), whereas extracts from zooplankton were antirachitic to a varying degree (15, 16). When this topic was investigated again about 30 years later, using better techniques, both phytoplankton and zooplankton were shown to contain several sterols that may function as pro-vitamins (21, 47). The main pro-vitamin of phytoplankton is pro-vitamin D_2, ergosterol, which can be converted to pre-vitamin D_2 by UV radiation (31, 32). Correspondingly, zooplankton contains mainly (75-80%) pro-vitamin D_3, 7-dehydrocholesterol, which is transformed to pre-vitamin D_3 after UV light exposure

(31). These observations therefore suggest that fish food is a source of vitamin D, provided that the plankton is exposed to sunlight. The general behavioural pattern of most zooplankton, in cold and temperate waters, is a seasonal as well as a diurnal vertical migration. They reside in the shallow water during spring and summer, and migrate to deeper waters during autumn and winter. Although the diurnal migration pattern makes them avoid the brightest light, they are still situated in the uppermost 10 m of the water mass, during morning and afternoon, and thus, expose themselves to UV light (35). In an attempt to calculate the balance between the ingested amount of vitamin D_3 and the liver vitamin D_3 content for the Atlantic cod (*Gadus morhua*), Bills (8) estimated that the cod had to eat 26 times its own weight in four weeks, a *reductio ad absurdum*, which made him suggest that fish might have an endogenous production of vitamin D_3.

A second possible way for fish to obtain their vitamin D_3 is thus through a nonphotochemical conversion of 7-dehydrocholesterol to pre-vitamin D_3. Juvenile channel catfish (*Ictalurus punctatus*) raised for nine weeks or six months in darkness and fed trimmed veal of very low vitamin D_3 content, showed no difference in liver vitamin D_3 content compared with the control group (8). Further studies on this topic were performed with Atlantic striped bass (*Roccus saxatilis*). Liver homogenates were incubated in the dark with ^{14}C-labelled 7-dehydrocholesterol, yielding low but significant concentrations of a radioactive substance with the properties of vitamin D_3 (10). Conversely, Sugisaki *et al.* (58) were unable to demonstrate a conversion of 7-dehydrocholesterol to more polar metabolites in the goldfish (*Carassius auratus*), suggesting that this species is unable to endogenously produce vitamin D_3 through a nonphotochemical pathway. A recent study by Norman and Norman (50) outlines four chemically possible ways for a nonphotochemical conversion of 7-dehydrocholesterol to vitamin D_3, but experimental evidence for the existence of these pathways is still lacking.

A third way for fish, as for mammals, to obtain vitamin D_3 is through a photochemical conversion of 7-dehydrocholesterol to vitamin D_3, in the skin, induced by UV radiation. Fish skin does contain pro-vitamin D_3 which is converted to vitamin D_3 after UV exposure. However, the levels of pro-vitamins are lower than in most mammals (31, 65) and UV light does not reach far in water, especially not in coastal seawater. Therefore, many fish species probably live below the penetration level of such light, and are thus unable to utilize this pathway for vitamin D_3 synthesis.

Metabolism of vitamin D_3 and circulating levels of vitamin D_3 metabolites in teleosts
The physiological importance of the vitamin D_3 system in teleosts has been extensively disputed over the years. Studies on channel catfish and rainbow trout (*Oncorhynchus mykiss*) have demonstrated vitamin D_3 to be a dietary requirement for normal growth and development (2, 6, 7, 11, 37, 40). Although some studies have been unable to demonstrate conversion of [3H]-vitamin D_3 and [3H]-25-hydroxycholecalciferol (25(OH)D_3) into more polar metabolites, and thus questioned the ability of teleosts to further metabolize vitamin D_3 (31, 32, 51, 69), 25-hydroxycholecalciferol-1-hydroxylase activity has been found in both liver and

kidney of several freshwater and seawater teleost species (28, 34, 66). Furthermore, rainbow trout and European eel (*Anguilla anguilla*) have been shown to metabolize $25(OH)D_3$ into several more polar metabolites, of which one co-chromatographed with 1,25-dihydroxyvitamin D_3 ($1,25(OH)_2D_3$) on HPLC and Sephadex LH 20 columns (4, 26). In contrast to mammals, the liver appears to be of major importance for both the 25- and the 1α-hydroxylation in fish. This is suggested by the higher 25-hydroxycholecalciferol-1-hydroxylase activity in the liver than in the kidney, and by the low circulating levels of $25(OH)D_3$ in fish (26, 60, 66).

With a few exceptions (7, 33), levels of circulating vitamin D_3, $25(OH)D_3$ and/or $1,25(OH)_2D_3$ have been observed in freshwater and seawater teleosts, elasmobranchs, as well as in the cyclostome, *Myxine glutinosa* (3, 5, 22, 31, 41, 44, 60, 66). The metabolite $24,25(OH)_2D_3$ has also been detected in the plasma of one teleost species, the yellowtail tuna (*Seriola lalandei* (66)). The physiological relevance of circulating vitamin D_3 metabolites is supported by reports of transport proteins for $25(OH)D_3$ in the plasma of several freshwater and marine teleost species (1, 25, 41, 60, 67). Although a specific binding of $25(OH)D_3$ to an α-globulin was first indicated (25), recent studies on several teleost species have all shown the major binding to be to a low-affinity lipoprotein. However, in carp (*Cyprinus carpio*) and rainbow trout, an additional high-affinity, low-capacity binding of $25(OH)D_3$ and $1,25(OH)_2D_3$ has been found (1, 60, 67, van Baelen and Sundell unpublished results). This high-affinity binding protein shows biochemical and physical characteristics very similar to the mammalian vitamin D binding protein (DBP (1, 67)).

Thus, the results presented to date demonstrate that both freshwater and marine teleosts have the ability to metabolize vitamin D_3 into more polar metabolites, and that they have circulating levels of three or four metabolites: vitamin D_3, $25(OH)D_3$, $1,25(OH)_2D_3$, and possibly also $24,25(OH)_2D_3$. However, since the circulating levels of $25(OH)D_3$, and probably also of $24,25(OH)_2D_3$, are lower in fish than in mammals, future studies on fish need highly sensitive techniques for the analysis of circulating metabolite levels.

Physiological effects of the vitamin D_3 system in teleosts

To date, there is substantial evidence for physiological effects of different vitamin D_3 metabolites on teleost calcium regulation, both *in vivo* and *in vitro*. Daily injections of vitamin D_3 and $1,25(OH)_2D_3$ induced a 20-30% hypercalcaemia (total plasma calcium levels) in the freshwater carp and catfish (*Heteropneustes fossilis;* irrespective of the water Ca levels being undetectable, 'normal' ($[Ca^{2+}]=0.8$ mM) or 'high' ($[Ca^{2+}]=14$ mM (55, 56, 62-64)). In freshwater American eel (*A. rostrata*) both vitamin D_3 and $1,25(OH)_2D_3$ increased total plasma calcium levels by 30% (19), in fed goldfish vitamin D_3 induced hypercalcaemia (17). Among the marine teleosts, only the Atlantic cod has been thoroughly investigated. $1,25(OH)_2D_3$ treatment *in vivo* induced hypercalcaemia by elevating the physiologically important ionized pool of calcium. The effect was present after 24 h and lasted for 5 days. None of the other metabolites tested in the study (vitamin D_3, $25(OH)D_3$, $24,25(OH)_2D_3$) affected the plasma calcium levels (61). Vitamin D_3 treatment of the Antarctic fish, *Pagothenia bernacchii*, increased both total and ionized plasma calcium levels, whereas injections of $1,25(OH)_2D_3$ decreased the ionized calcium concentration without altering the total plasma calcium levels, suggesting an increased binding of calcium to plasma proteins in this species (20).

Effects of vitamin D_3 and its metabolites on calcium fluxes across different target organs *in vitro* are also evident. Bone mineralization is decreased by injections of $1,25(OH)_2D_3$ in the freshwater-adapted eel and tilapia (*Sarotherodon mossambicus* (38, 39, 68)). Conversely, in the sexually mature female eel in seawater, this metabolite converts lining cells into osteoblasts and increases osteoblastic activity. Mature eels normally have a higher bone demineralization rate than immature eels, probably due to an increased demand for calcium, which binds to vitellogenin during sexual maturation (9, 39). Injections of $1,25(OH)_2D_3$ in mature female eels prevented this demineralization, thus acting, in this developmental stage, in a similar way to $24,25(OH)_2D_3$, which increased osteoblastic activity as reported for tilapia (68). To date, only one study has dealt with the effects of vitamin D_3 on the teleost kidney, and this was limited to renal phosphate handling, demonstrating that vitamin D_3 increased the tubular reabsorption of phosphate in the American eel (18). Data concerning the effects of vitamin D_3 metabolites on calcium fluxes across kidney and gills are lacking, and represents a research area where future studies are greatly needed. The intestine has received more attention in this context, and responds to several different metabolites of the vitamin D_3 system. This will be dealt with in more detail in the next part of this chapter.

Intestinal calcium regulation of the vitamin D_3 system: genomic versus non-genomic actions

Endocrine calcium regulation was probably modified during the water-to-land transition, giving rise to new endocrine systems and new roles for already present hormones. However, it can be hypothesized that the evolution of the vitamin D_3 system has followed a different pattern. The vitamin D_3 system seems to have evolved similarly in terrestrial vertebrates and freshwater fish, both exposed to what we can consider a hypocalcic environment. Marine teleosts, on the other hand, live in a hypercalcic environment and the evolution of the vitamin D_3 system appears to have diverged in this group. This hypothesis will be further clarified below.

It is well established that the vitamin D_3 metabolite most active in the regulation of the calcium homeostasis of mammals and birds is $1,25(OH)_2D_3$. This hormone acts in two different ways on intestinal calcium uptake.

The first route is via a classic, genomic pathway, in which $1,25(OH)_2D_3$ binds to a tissue-specific receptor and the receptor-hormone complex subsequently interacts with the DNA. This modulates transcription of mRNA encoding for specific proteins, bringing about an increase in the intestinal calcium uptake after 10 to 24 h (23).

The second route is via a non-genomic pathway, in which $1,25(OH)_2D_3$ probably interacts with a membrane-bound receptor and affects different processes within the cell: stimulation of phospholipid metabolism, release of lysosomal enzymes, stimulation of alkaline phosphatases and activation of calcium channels (see also (46)).

Through this non-genomic mechanism, $1,25(OH)_2D_3$ increases the calcium uptake within 2-10 min (45).

In teleosts, only a few recent studies have attempted to separate the effects of the vitamin D_3 metabolites according to the two above pathways. However, there is a line of evidence for the presence of the genomic pathway in both marine and freshwater teleosts. Treatment with $1,25(OH)_2D_3$ has induced hypercalcaemia in all freshwater and seawater fish examined, with the exception of *P. bernacchii* (20). In the Atlantic cod, the $1,25(OH)_2D_3$-induced hypercalcaemia is apparent after 24 h but not after 12 h, indicating a slow and thus genome-mediated effect (61). In the chick, one of the best studied models for the genomic effects of $1,25(OH)_2D_3$ on intestinal calcium handling, a single dose of $1,25(OH)_2D_3$ results in a maximal mRNA production after 4-6 h and a maximal biological response, i.e. stimulated intestinal calcium uptake, after 10-12 h (48). The longer time needed for $1,25(OH)_2D_3$ to exert the genomic effect in the cod than in the chick is probably due to the difference in body temperature (cf. (61)). Further evidence for the genome-mediated effect of $1,25(OH)_2D_3$ in both freshwater and marine teleosts, is the presence of high-affinity, low-capacity binding of $1,25(OH)_2D_3$ in high-salt cytosol preparations from the intestines of both the freshwater-adapted eel and the marine cod (41, 60). This specific binder shows biochemical and physical characteristics with striking similarity to the well defined n-VDR, the chick nuclear $1,25(OH)_2D_3$-receptor (50, 60). A classic response of enterocytes to $1,25(OH)_2D_3$ treatment is the production of a calcium binding protein (calbindin D-28k, which is found in avian, reptile and amphibian intestines, and calbindin D-9k, which is found in mammalian intestine (54)). Several studies have attempted to demonstrate the existence of calbindins in different teleost species and tissues. Antibodies against chick calbindin D-28k have not cross-reacted with fish tissues except for those from the brain (12, 53). This is possibly due to sequence dissimilarities among the chick and the fish calbindins, especially since indirect evidence for calcium binding proteins, using a ^{45}Ca binding procedure, has been found in different fish tissues, including the intestine of several both freshwater and marine teleost species (13, 27, 52).

Thus, whereas a genome-mediated effect has been demonstrated both in vertebrates in hypocalcic (freshwater and terrestrial) and hypercalcic (marine) environments, the principal difference appears to be in the evolution of the nongenomic effect. Transcaltachia, defined as the rapid, nongenomic $1,25(OH)_2D_3$-mediated stimulation of intestinal calcium uptake, seems only to be present in freshwater teleosts. The strongest proof for the nongenomic effect in terrestrial animals has been obtained using a vascularly perfused duodenal loop of vitamin-D replete chicks. Employing this method, an increased calcium uptake was observed after 2 min of perfusion with as low concentrations of $1,25(OH)_2D_3$ as 130 pM (45). A similar technique has been used in two teleost species: the freshwater-adapted European eel and the marine Atlantic cod. Perfusion with $1,25(OH)_2D_3$ increased the intestinal calcium uptake within 5 min in the eel, demonstrating the presence of transcaltachia (14). The same rapid response was seen also during perfusion with the precursor molecule vitamin D_3, suggesting that the nongenome-mediated response is less differentiated in freshwater fish than in mammals and birds. In the Atlantic cod, on the other hand, perfusion studies using both a pharmacological concentration of $1,25(OH)_2D_3$ (24 nM (59)) and a series of more physiologically relevant doses (85, 128, 256 and 640 pM compared with the circulating levels of 128 ± 38 pM (36, 61)) of the same metabolite, did not reveal a rapid effect on the intestinal calcium uptake. Thus, no transcaltachia is present in this marine teleost.

Instead, the metabolite $24,25(OH)_2D_3$ was discovered to rapidly affect the intestinal calcium uptake in the cod. Perfusions with $24,25(OH)_2D_3$ at concentrations ranging from 2-20 nM (circulating levels of $24,25(OH)_2D_3$ have only been reported for the yellowtail tuna as being 2 nM (66)) decreased the calcium uptake in a dose-dependent manner within 10 min (36, 59). Thus, in this marine species, $24,25(OH)_2D_3$ rather than $1,25(OH)_2D_3$ seems to exert its effect *via* a rapid nongenomic pathway by decreasing the intestinal calcium uptake.

References

1   Allewaert K, van Baelen H & Bouillon R (1988) Two 25-hydroxycholecalciferol binding proteins in carp serum. *Vitamin D: Molecular, Cellular and Clinical Endocrinology,* pp 692-693. Eds AW Norman, K Schaefer, HG Grigoleit & D von Herrath. Berlin, New York: Walter de Gruyter.

2   Andrews JW, Murai T & Page JW (1980) Effects of dietary cholecalciferol and ergocalciferol on catfish. *Aquaculture* **19** 49-54.

3   Avioli LV, Sonn Y, Jo D, Nahn TH, Haussler MR & Chandler JS (1981) 1,25-dihydroxyvitamin D in male, nonspawning female, and spawning female trout. *Proceedings of the Society for Experimental Biology and Medicine* **166** 291-293.

4   Bailly du Bois M, Milet C, Garabedian M, Guillozo H, Martelly E, Lopez E & Balsan S (1988) Calcium-dependent metabolism of 25-hydroxycholecalciferol in silver eel tissues. *General and Comparative Endocrinology* **71** 1-9.

5   Barnett BJ, Cho CY & Slinger SJ (1979) The essentiality of cholecalciferol in the diets of rainbow trout (*Salmo gairdneri*). *Comparative Biochemistry and Physiology* **63A** 291-297.

6   Barnett BJ, Cho CY & Slinger SJ (1982) Relative biopotency of dietary ergocalciferol and cholecalciferol and the role of and requirement for vitamin D in rainbow trout (*Salmo gairdneri*). *Journal of Nutrition* **112** 2011-2019.

7   Barnett BJ, Jones G, Cho CY & Slinger SJ (1982) The biological activity of 25-hydroxycholecalciferol and 1,25- dihydroxycholecalciferol for rainbow trout (*Salmo gairdneri*). *Journal of Nutrition* **112** 2020-2026.

8   Bills CE (1927) The distribution of vitamin D, with some notes on its possible origin. *Journal of Biological Chemistry* **72** 751-758.

9   Björnsson BTh, Haux C, Förlin L & Deftos LJ (1986) The involvement of calcitonin in the reproductive physiology of the rainbow trout. *Journal of Endocrinology* **108** 17-23.

10   Blondin GA, Kulkarni BD & Nes WR (1967) A study of the origin of vitamin-D from 7-dehydrocholesterol in fish. *Comparative Biochemistry and Physiology* **20** 379-390.

11   Brown PB & Robinson EH (1992) Vitamin D studies with channel catfish (*Ictalurus punctatus*) reared in calcium-free water. *Comparative Biochemistry and Physiology* **103A** 213-219.

12   Celio MR, Baier W, Schärer L, Gregersen HJ, de Viragh PA & Norman AW (1990) Monoclonal antibodies directed against the calcium binding protein Calbindin D-28k. *Cell Calcium* **11** 599-602.

13   Chartier Baraduc MM (1973) Présence et thermostabilité de protéines liant le calcium dans les muqueuses intestinales et branchiales de divers Téléostéens. *Comptes Rendus de l' Academie des Sciences, Paris* **276D** 785-788.

14   Chartier MM, Millet C, Martelly E, Lopez E & Warrot S (1979) Stimulation par la vitamine D_3 et le 1,25-dihydroxyvitamine D_3 de l'absorption intestinale du calcium chez l'anguille (*Anguilla anguilla* L.). *Journal de Physiologie, Paris* **75** 275-282.

15   Copping AM (1934) Origin of vitamin D in cod-liver oil: vitamin D content of zooplankton. *The Biochemical Journal* **28** 1516-1520.

16   Drummond JC & Gunther ER (1930) Vitamin content of marine plankton. *Nature* **126** 398.

17   Fenwick JC (1984) Effect of vitamin D_3 (cholecalciferol) on plasma calcium and intestinal [45]calcium absorption in goldfish, *Carrasius auratus L. Canadian Journal of Zoology* **62** 34-36.

18   Fenwick JC & Vermette MG (1989) Vitamin D_3 and the renal handling of phosphate in American eels. *Fish Physiology and Biochemistry* **7** 351-358.

19   Fenwick JC, Smith K, Smith J & Flik G (1984) Effect of various vitamin D analogs on plasma calcium and phosphorus and intestinal calcium absorption in fed and unfed american eels, *Anguilla rostrata. General and Comparative Endocrinology* **55** 398-404.

20   Fenwick JC, Davison W & Forster ME (1994) *In vivo* calcitropic effect of some vitamin D compounds in the marine Antarctic teleost, *Pagothenia bernacchii. Fish Physiology and Biochemistry* **12** 479-484.

21   Fieser LF & Fieser M (1959) *Steroids.* London: Chapman & Hall Ltd

22   Glowacki J, Deftos LJ, Mayer E, Norman AW & Henry H (1982) Chondrichthyes cannot resorb implanted bone and have calcium-regulating hormones. *Vitamin D, Chemical, Biochemical and Clinical Endocrinology of Calcium Metabolism,* pp 613-615. Berlin: Walter de Gruyter.

23   Hannah SS & Norman AW (1994) $1\alpha,25(OH)_2$ vitamin D_3-regulated expression of the eukaryotic genome. *Nutrition Reviews* **52** 376-382.

24   Haussler MR, Myrtle JF & Norman AW (1968) The association of a metabolite of vitamin D_3 with intestinal mucosa chromatin *in vivo. Journal of Biological Chemistry* **243** 4055-4064.

25   Hay AWM & Watson G (1976) The plasma transport proteins of 25-hydroxycholecalciferol in mammals. *Comparative Biochemistry and Physiology* **53B** 163-166.

26   Hayes ME, Guilland-Cumming DF, Russell RGG & Henderson IW (1986) Metabolism of 25-hydroxycholecalciferol in a teleost fish, the rainbow trout (*Salmo gairdneri*). *General and Comparative Endocrinology* **64** 143-150.

27   Hearn PR, Tomlinson S, Mellersh H, Preston CJ, Kenyon CJ & Russell RGG (1978) Low molecular weight calcium-binding protein from the kidney and gill of the freshwater eel (*Anguilla anguilla*). *Journal of Endocrinology* **79** 36P-37P.

28   Henry HL & Norman AW (1975) Presence of renal 25-hydroxyvitamin-D-1-hydroxylase in species of all vertebrate classes. *Comparative Biochemistry and Physiology* **50B** 431-434.

29   Henry HL & Norman AW (1992) Disorders of bone and mineral metabolism. *Metabolism of vitamin D,* pp 149-163. Eds FL Coe & MJ Favus. New York: Raven Press.

30   Hess AF & Weinstock M (1924) Antirachitic properties imparted to inert fluids and to green vegetables by ultra-violet irradiation. *Journal of Biological Chemistry* **62** 301-313.

31   Holick MF (1989) Phylogenetic and evolutionary aspects of vitamin D from phytoplankton to humans. *Vertebrate Endocrinology: Fundamentals and Biomedical Implications,* vol 3, pp 7-43. Eds PKT Pang & MP Schreibman. New York: Academic Press.

32   Holick MF, Holick SA & Guillard RL (1982) On the origin and metabolism of vitamin D in the sea. *Comparative Endocrinology of Calcium Regulation,* pp 85-91. Eds C Oguro & PKT Pang. Tokyo: Japan Scientific Press.

33   Hollis BW, Burton JH & Draper HH (1977) A binding assay for 25-hydroxycalciferols and 24R,25-dihydroxycalciferols using bovine plasma globulin. *Steroids* **30** 285-293.

34   Kenny AD, Baksi SN, Galli-gallardo SM & Pang PKT (1977) Vitamin D metabolism in amphibia and fish. *Federation Proceedings* **36** 1097.

35   Lalli CM & Parsons TR (1993) *Biological Oceanography: An Introduction.* Oxford: Pergamon Press.

36   Larsson D, Björnsson BTh & Sundell K (1995) Physiological concentrations of 24,25-dihydroxyvitamin D_3 rapidly decease the *in vitro* intestinal calcium uptake in the Atlantic cod, *Gadus morhua. General and Comparative Endocrinology* **100** 211-217.

37   Leatherland JF, Barnett BJ, Cho CY & Slinger SJ (1980) Effect of dietary cholecalciferol deficiency on plasma thyroid hormone concentrations in rainbow trout, *Salmo gairdneri* (Pisces, Salmonidae). *Environmental Biology of Fishes* **5** 167-173.

38   Lopez E, Peignoux-Deville J, Lallier F, Colston KW & MacIntyre I (1977) Responses of bone metabolism in the eel (*Anguilla anguilla*) to injections of 1,25-dihydroxyvitamin D_3. *Calcified Tissue Research* **22** 19-23.

39   Lopez E, MacIntyre I, Martelly E, Lallier F & Vidal B (1980) Paradoxical effect of 1,25 dihydroxycholecalciferol on osteoblastic and osteoclastic activity in the skeleton of the eel *Anguilla anguilla* L. *Calcified Tissue International* **32** 83-87.

40   Lovell RT & Li YP (1978) Essentiality of vitamin D in diets of channel catfish (*Ictalurus punctatus*). *Transactions of the American Fisheries Society* **107** 809-811.

41   Marcocci C, Freake HC, Iwasaki J, Lopez E & MacIntyre I (1982) Demonstration and organ distribution of the 1,25-dihydroxyvitamin D_3-binding protein in fish (*A. anguilla*). *Endocrinology* **110** 1347-1354.

42   McCollum EV, Simmonds N, Becker JE & Shipley PG (1922) Studies on experimental rickets. An experimental demonstration of existence of a vitamin which promotes calcium deposition. *Journal of Biological Chemistry* **53** 293-312.

43   Mellanby E (1919) An experimental investigation on rickets. *Lancet* **i** 407-412.

44   Nahm TH, Lee SW, Fausto A, Sonn Y & Avioli LV (1979) 250HD, a circulating vitamin D metabolite in fish. *Biochemical and Biophysical Research Communications* **82** 396-402.

45   Nemere I & Norman AW (1986) Parathyroid hormone stimulates calcium transport in perfused duodena from normal chicks: comparison with the rapid (Transcaltachic) effect of 1,25-dihydroxyvitamin D_3. *Endocrinology* **119** 1406-1408.

46   Nemere I, Zhou LX & Norman AW (1993) Nontranscriptional effects of steroid hormones. *Receptor* **3** 277-291.

47   Nes WR & McKean ML (1977) *Biochemistry of Steroids and Other Isopentenoids.* Baltimore: University Park Press.

48   Norman AW (1990) Intestinal calcium absorption: a vitamin D-hormone-mediated adaptive response. *American Journal for Clinical Nutrition* **51** 290-300.

49   Norman TC & Norman AW (1993) Consideration of chemical mechanisms for the nonphotochemical production of vitamin D_3 in biological systems. *Bioorganic and Medicinal Chemistry Letters* **3** 1785-1788.

50   Norman AW, Roth J & Orci L (1982) The vitamin D endocrine system: steroid metabolism, hormone receptors, and biological response (Calcium Binding Proteins). *Endocrine Reviews* **3** 331-366.

51   Oizumi K & Monder C (1972) Localization and metabolism of $1,2\text{-}^3$H-vitamin D_3 and $26,27\text{-}^3$H-25-hydroxycholecalciferol in goldfish (*Carassius auratus*). *Comparative Biochemistry Physiology* **42B** 523-532.

52   Ooizumi K, Moriuchi S & Hosoya N (1970) Comparative study of vitamin-D_3-induced calcium binding protein. *Vitamins* **42** 171-175.

53   Parmentier M, Ghysens M, Rypens F, Lawson DEM, Pasteels JL & Pochet R (1987) Calbindin in vertebrate classes: immunohistochemical localization and western blot analysis. *General and Comparative Endocrinology* **65** 399-407.

54   Robertson DR (1993) Comparative aspects of intestinal calcium transport in fish and amphibians. *Zoological Science* **10** 223-234.

55   Srivastav AK & Singh S (1992) Effect of vitamin D_3 on serum calcium and inorganic phosphate levels of the freshwater catfish, *Heteropneustes fossilis*, maintained in artificial freshwater, calcium-rich freshwater, and calcium deficient freshwater. *General and Comparative Endocrinology* **87** 63-70.

56   Srivastav SK, Jaiswal R & Srivastav AK (1993) Response of serum calcium to administration of 1,25-dihydroxyvitamin D_3 in the freshwater carp *Cyprinus carpio* maintained either in artificial freshwater, calcium-rich freshwater or calcium-deficient freshwater. *Acta Physiologica Hungarica* **81** 269- 275.

57   Steenbock H & Black A (1924) The induction of growth-promoting and calcifying properties in a ration by exposure to ultra-violet light. *Journal of Biological Chemistry* **61** 405-422.

58   Sugisaki N, Welcher M & Monder C (1974) Lack of vitamin D_3 synthesis by goldfish (*Carassius auratus L.*). *Comparative Biochemistry and Physiology* **49B** 647-653.

59   Sundell K & Björnsson BTh (1990) Effects of vitamin D_3, 25(OH) vitamin D_3, $24,25(OH)_2$ vitamin D_3, and $1,25(OH)_2$ vitamin D_3 on the *in vitro* intestinal calcium absorption in the marine teleost, Atlantic cod (*Gadus morhua*). *General and Comparative Endocrinology* **78** 74-79.

60   Sundell K, Bishop JE, Björnsson BTh & Norman AW (1992) 1,25-dihydroxyvitamin D_3 in the Atlantic cod: plasma levels, a plasma binding component, and organ distribution of a high affinity receptor. *Endocrinology* **131** 2279-2286.

61   Sundell K, Norman AW & Björnsson BTh (1993) $1,25(OH)_2$ vitamin D_3 increases ionized plasma calcium concentrations in the immature Atlantic cod *Gadus morhua. General and Comparative Endocrinology* **91** 344-351.

62   Swarup K & Srivastav SP (1982) Vitamin D_3-induced hypercalcemia in male catfish, *Clarias batrachus. General and Comparative Endocrinology* **46** 271-274.

63   Swarup K, Norman AW, Srivastav AK & Srivastav SP (1984) Dose-dependent vitamin D_3 and 1,25-dihydroxyvitamin D_3-induced hypercalcemia and hyperphosphatemia in male catfish *Clarias batrachus. Comparative Biochemistry and Physiology* **78B** 553-555.

64   Swarup K, Das VK & Norman AW (1991) Dose-dependent vitamin D_3 and 1,25-dihydroxyvitamin D_3-induced hypercalcemia and hyperphosphatemia in male cyprinoid *Cyprinus carpio. Comparative Biochemistry and Physiology* **100A** 445-447.

65   Takeuchi A, Okano T, Sayamoto M, Sawamura S, Kobayashi T, Motosugi M & Yamakawa T (1986) Tissue distribution of 7-dehydrocholesterol, vitamin D_3 and 25-hydroxyvitamin D_3 in several species of fish. *Journal of Nutritional Science and Vitaminology* **32** 13-22.

66   Takeuchi A, Okano T & Kobayashi T (1991) The existence of 25-hydroxyvitamin D_3-1α-hydroxylase in the liver of carp and bastard halibut. *Life Sciences* **48** 275-282.

67   Van Baelen H, Allewaert K & Bouillon R (1988) New aspects of the plasma carrier protein for 25-hydroxycholecalciferol in vertebrates. *Annuals of New York Academic Sciences* **538** 60-68.

68   Wendelaar Bonga SE, Lammers PI & van der Meij JCA (1983) Effects of 1,25-and 24,25-dihydroxyvitamin D_3 on bone formation in the chichlid teleost *Sarotherodon mossambicus. Cell Tissue Research* **228** 117-126.

69   Yanda DM & Ghazarian JG (1981) Vitamin D and 25-hydroxyvitamin D in rainbow trout (*Salmo gairdneri*): cytochrome P-450 and biotransformations of the vitamins. *Comparative Biochemistry and Physiology* **69B** 183-188.

The Comparative Endocrinology of Calcium Regulation
Eds C Dacke, J Danks, I Caple & G Flik, pp 85-90
Journal of Endocrinology Ltd, Bristol (1996)

Parathyroid hormone-related protein in lower aquatic vertebrates

M K Trivett[1,2], R A Officer[2], J G Clement[3], T I Walker[4],
J M Joss[5], P M Ingleton[6], T J Martin[1] and J A Danks[1]

[1]St Vincent's Institute of Medical Research, St Vincent's Hospital, 41
Victoria Parade, Fitzroy, 3065, Australia, [2]Department of Zoology and
[3]School of Dental Science, University of Melbourne, Parkville, 3052,
Australia, [4]Victorian Fisheries Research Institute, Queenscliff, 3225,
Australia, [5]School of Biological Science, Macquarie University, Sydney
2109, Australia and [6]Institutes of Endocrinology and Cancer Studies,
The Medical School, Beech Hill Road, Sheffield S10 2RX, UK

Introduction

Parathyroid hormone-related protein (PTHrP) first came to attention as the mediator of the humoral hypercalcaemia of malignancy (HHM). It is found in squamous cell carcinomas, regardless of origin (5) but has also been found in normal human skin and occurs in high levels during pregnancy where it has a role in calcium transport between the mother and fetus (15). PTHrP causes elevations in plasma calcium by mobilizing calcium from bone and by inhibiting phosphate metabolism in the kidney. It has N-terminal homology with parathyroid hormone (PTH), with eight of the first 13 amino acids being identical. In tetrapods, PTH is produced by the parathyroid glands and is the major hypercalcaemic factor in this group of animals, acting on the bone and kidney. Calcium and phosphate are acquired through the diet, and bone acts as a reservoir for these ions. The tetrapods have a closed system of calcium regulation (4) with calcium levels being regulated through the actions of PTH, vitamin D_3 and calcitonin. Hormones known to affect calcium regulation in tetrapods have been observed in lower aquatic vertebrates, however, the role of these factors in this group of animals remains unclear. In contrast to the tetrapods, lower aquatic vertebrates lack encapsulated parathyroid glands and many live in a calcium-rich environment. PTHrP has been demonstrated immunohistochemically in the pituitary of the sea bream, *Sparus aurata*, a marine teleost (6) and circulating levels in the plasma of this fish were comparable to those of patients with HHM. It appears that the site of production of PTHrP was the pituitary of the fish, and it was suggested that in some species, PTHrP may act as a classical hormone. Bony fish also have vitamin D_3 (in their livers), calcitonin and stanniocalcin, which may have roles in calcium regulation. Recently, PTHrP was observed in the tissues and plasma of the dogfish, *Scyliorhinus canicula* (12), a cartilaginous fish. The chondrichthyans have cartilaginous skeletons which have only pockets of calcified tissue (18) and it is unclear whether this skeleton has a role in calcium regulation in the animal. Like the bony fish, cartilaginous fish do not have PTH but unlike the bony fish they lack stanniocalcin and the levels of vitamin D_3 in their livers are negligible (3). They do have levels of calcitonin similar to those of bony fish or mammals, but the calcitonin does not appear to have a role in calcium regulation (4). The cartilaginous fish have an open system of calcium regulation and the problem faced by the animal is to prevent calcium influx from the external environment. The target organs for calcium regulation in these animals are likely to be the gut, gills, skin and nephron (4). The cyclostomes (lampreys and hagfish), living representatives of the oldest vertebrates, lack PTH, vitamin D_3 metabolites and calcitonin but control plasma calcium to some extent by exchange of ions with the water in specialized epithelia (3). In this paper, the tissues of lower aquatic vertebrates were examined by immunohistochemistry to investigate the evolutionary history of PTHrP and its distribution in the lower aquatic vertebrates.

Materials and methods

Tissues
Skin and kidney samples from the upstream adult migrant of the lamprey *Geotria australis* were formalin fixed, routinely processed and embedded in paraffin blocks (gift of Glen Power, Murdoch University, Murdoch, Western Australia). Two whole larval lungfish of *Neoceratodus forsteri* (kindly donated by Jean Joss, Macquarie University, Sydney, NSW) were bisected after fixation in formalin, routinely processed, embedded in paraffin wax and serially sectioned. Fresh tissue was collected from four species of elasmobranchs, held at the Victorian Fisheries Research Institute (Queenscliff, Victoria, Australia). Sharks were anaesthetized in MS-222 and decapitated. Pieces of dorsal and ventral skins, as well as kidney and liver samples were taken from ten gummy sharks, *Mustelus antarcticus*, two school sharks, *Galeorhinus galeus*, four banjo sharks (Southern Fiddler rays), *Trygonorrhina fasciata* and two common spotted stingarees, *Urolophus gigas*. Tissues were fixed in either formalin (for less than 24 h) or in sublimated Bouin-Hollande (48 to 72 h). Duplicate samples of dorsal skin from each fixative were decalcified in formic and citric acid. Tissues fixed in sublimated Bouin-Hollande were washed in running tap water for half a day to remove picric acid and transferred to 70% alcohol before further processing. After routine processing the tissues were embedded in paraffin wax.

Immunohistochemistry
Sections (5 μm) for PTHrP immunohistochemistry were mounted on glass microscope slides, coated with 2% 3-aminopropyltriethoxysilane (Sigma, St Louis, MO) in acetone. The following antibodies raised to human PTHrP were used: (i) antibody to PTHrP(1-16) raised in New Zealand white rabbits (R1133) and (ii) antibody to PTHrP(50-69) raised in sheep (S8094). Staining was performed according to the method of Danks *et al.* (5), using a standard immunoperoxidase technique (17). Tissues fixed in sublimated Bouin-Hollande had mercury precipitate removed by washing in 1% iodine in 70% alcohol (1 min), followed by rinsing in 5% sodium thiosulphate (1 min). Sections were then washed in running tap water for half an hour. Briefly, endogenous peroxidase activity was blocked in 1% hydrogen peroxide,

followed by blocking of non-specific binding in either 10% swine serum (R1133) or 10% rabbit serum (S8094). Sections were incubated in the primary antiserum, followed by incubation in a secondary antibody. The antigen was visualized by addition of PAP complex, using 3,3'-diaminobenzidine (DAB) as the chromogen. Sections were counterstained in Harris' haematoxylin.

Controls

In each experiment, normal human skin was included as a positive control, while non-immune rabbit serum was a negative control. Other controls included omission of the various antibody layers (primary antiserum, secondary antibody and the PAP complex) and the use of an unrelated rabbit polyclonal antibody and a rabbit polyclonal antibody to human PTH(1-34) (BioGenex, San Raman, CA) to confirm the specificity of PTHrP staining.

Results

In the developing lungfish, PTHrP was localized to the epidermis of dorsal and ventral skin as well as the underlying muscle, the kidney tubules, the notochord and the developing brain. Figure 1a shows the dorsal epidermis of larval lungfish stained with antibody to PTHrP(1-16). There is cytoplasmic staining of the cells in the epidermis, the underlying dermis being negative. The developing brain is also positive and is visible in the bottom left of the picture. Figure 1b is the adjacent section where non-immune rabbit serum was substituted for the primary antibody. Antigen was found in kidney tubules, the lining of the urinary ducts, skeletal muscle and epidermis of the dorsal and ventral skins in all of the sharks and rays studied. The cells in the epidermis adjacent to the denticles on the dorsal skin of gummy shark were positive (Fig. 2a), but negative when non-immune rabbit serum was used (Fig. 2b). PTHrP was not observed in the glomeruli of the kidney, nor in the liver. In the lamprey, PTHrP was found in the nuclei of kidney tubules, but not in the haemopoietic tissue (Figs 3a and 3b), and in the epidermis of the dorsal skin, but not in the underlying dermis. The pattern of staining was similar with both of the PTHrP antibodies. Deletion of primary antiserum, the secondary antibody or the PAP complex abolished the staining. No staining was observed in the regions to which PTHrP was localized when an unrelated rabbit polyclonal antibody or the PTH(1-84) antibody were used.

Discussion

In all of the lower aquatic vertebrates studied, PTHrP was localized to the kidney tubules and the skin, which is the pattern seen in the tetrapod vertebrates, indicating that there is conservation of the tissue distribution of PTHrP in the vertebrates. The N-terminal and mid-molecule antibodies localized PTHrP to the same areas, suggesting that there is conservation of the amino acid sequence from humans through to the lamprey and that there may be homology between the PTHrP molecules found in humans and fish. The amino acid sequences determined for tetrapod vertebrates show high conservation from residues 1 to 111, with divergence in the C-terminal region of the PTHrP molecule. The presence of PTHrP in the lamprey suggests that PTHrP is of

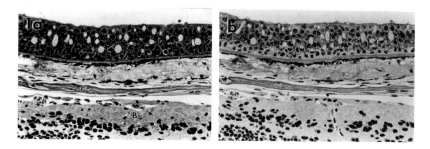

Fig. 1 (a) Section of whole larval lungfish stained with antibody to the N-terminus of PTHrP (R1133). The dorsal skin is positive, as is the developing brain. Note that the line of cartilage above the brain is negative for PTHrP. (b) Section of whole larval lungfish with non-immune rabbit serum.

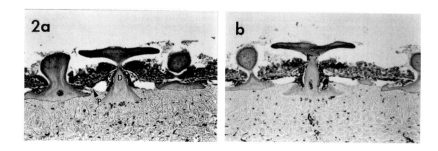

Fig. 2 (a) Dorsal skin of a gummy shark stained for PTHrP (R1133). The epidermis adjacent to the denticles is positive for PTHrP. A layer of dark pigment may also be seen. (b) Dorsal skin of shark with non-immune rabbit serum.

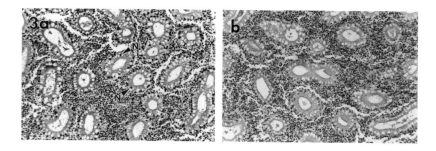

Fig. 3 (a) Lamprey kidney tubules stained with antibody to the N-terminus of PTHrP (R1133). The nuclei of the tubules and the tubules themselves are positive, but tissue surrounding the tubules is negative. (b) Lamprey kidney section with antibody substituted with non-immune rabbit serum.

ancient origin, and its widespread distribution among the vertebrates and conservation of localization raise the question of whether PTHrP has a fundamental physiological role in the vertebrates, perhaps in calcium regulation, or the regulation of other ions. The localization of PTHrP to the nuclei of the tubules in the lamprey kidney corroborates observations by Karaplis (14), who observed PTHrP in the nucleoli of transfected Cos-7 cells, while investigating the possibility that PTHrP could have intracellular actions. The sites in which PTHrP was found in this survey of lower aquatic vertebrates indicate that it could be involved in ion regulation. The parathyroids in tetrapods play a major role in calcium regulation, but in the lower aquatic vertebrates the pituitary assumes a greater role (2). It is possible that PTHrP and PTH evolved from a common ancestral gene after a duplication event, but the role that PTHrP may have in normal human adult calcium homeostasis has still not been determined.

Parsons *et al.* (16) suggested that a hypercalcaemic factor exists in the pituitary gland of fish that is immunologically similar to mammalian PTH. Pituitary prolactin controls calcium balance and integumental permeability to water and ions in freshwater fish (8, 11). These actions of prolactin are slow (it takes days to develop) and may be restricted to fish living in freshwater (7). Studies by Ball *et al.* (1) support the hypothesis of a second hypercalcaemic factor other than prolactin existing in the pituitary of fish. Kaneko and Pang (13) observed PTH-like immunoreactivity in the goldfish brain, but a hypercalcaemic function was not demonstrated, and goldfish plasma was also found to have PTH-like immunoreactivity (10). PTHrP has been localized to the pituitary of the Coho salmon by HPLC (9) and in the sea bream by immunohistochemistry (6). Tissues of the dogfish were also found to have PTHrP (12) in sites that could function in the regulation of ions. The regulation of calcium in fish is poorly understood and PTHrP may play an important role or have as yet undetermined roles in normal fish physiology. Understanding the role of PTHrP in fish may give insight into its role in the normal physiology of mammals and other tetrapods.

References

1   Ball JN, Uchiyama M and Pang PKT (1982) Pituitary responses to calcium deprivation in a euryhaline teleost, *Fundulus heteoclitus*, adapted to artificial seawater. *General and Comparative Endocrinology* **46** 480-485.

2   Clark N (1983) Evolution of calcium regulation in lower vertebrates. *American Zoologist* **23** 719-727.

3   Clement JG (1992) Re-examination of the fine structure of endoskeletal mineralization in chondrichthyes: implications for growth, ageing and calcium homeostasis. *Australian Journal of Marine and Freshwater Research* **43** 157-181.

4   Dacke CG (1979) *Calcium Regulation in Sub-Mammalian Vertebrates.* London: Academic Press.

5   Danks JA, Ebeling PR, Hayman J, Chou ST, Moseley JM, Dunlop J, Kemp BE and Martin TJ (1989) Parathyroid hormone-related protein: immunohistochemical localization in cancers and normal skin. *Journal of Bone and Mineral Research* **4** 273-278.

6   Danks JA, Devlin AJ, Ho PMW, Diefenbach-Jagger H, Power DM, Canario A, Martin TJ and Ingleton PM (1993) Parathyroid hormone-related protein is a factor in normal fish pituitary. *General and Comparative Endocrinology* **92** 201-212.

7   Fargher RC and McKeown BA (1989) The effect of prolactin on calcium homeostasis in Coho salmon (*Oncorhynchus kisutch*). *General and Comparative Endocrinology* **73** 398-403.

8   Flik G, Rentier-Delrue F and Wendelaar Bonga SE (1994) Calcitropic effects of recombinant prolactins in *Oreochromis mossambicus*. *American Journal of Physiology* **266** R1302-R1308.

9   Fraser RA, Kaneko T, Pang PKT and Harvey S (1991) Hypo- and hypercalcemic peptides in fish pituitary glands. *American Journal of Physiology* **260** R622-R626.

10   Harvey S, Zeng Y-Y and Pang PKT (1987) Parathyroid hormone-like immunoreactivity in fish plasma tissues. *General and Comparative Endocrinology* **68** 136-146.

11   Hirano T, Ogasawara T, Bolton JP, Collie NL, Hasgawa S and Iwata M (1981) Osmoregulatory role of prolactin in lower vertebrates. *Comparative Physiology of Environmental Adaptations,* pp 112-124. Eds R Kirsch and B Lahlou. Basel: Karger.

12   Ingleton PM, Hazon N, Ho PMW, Martin TJ and Danks JA (1995) Immunodetection of parathyroid hormone-related protein in plasma and tissues of an elasmobranch (*Scyliorhinus canicula*). *General and Comparative Endocrinology* **98** 211-218.

13   Kaneko T and Pang PKT (1987) Immunocytochemical detection of a parathyroid hormone-like substance in the goldfish brain and pituitary gland. *General and Comparative Endocrinology* **68** 147-152.

14   Karaplis AC, Henderson JE, Amizuka N, Lanske BKM, Biasotto D and Goltzman D (1994) Identification of a functional nucleolar localization signal in parathyroid hormone-related protein. *Journal of Bone and Mineral Research* **9** (Suppl 1) S121.

15   Martin TJ, Moseley JM and Gillespie MT (1991) Parathyroid hormone-related protein: biochemistry and molecular biology. *Critical Reviews in Biochemistry and Molecular Biology* **36** 377-395.

16   Parsons JA, Gray D, Rafferty B and Zanelli J (1979) Evidence for a hypercalcemic factor in the fish pituitary immunologically related to mammalian parathyroid hormone. *Endocrinology of Calcium Metabolism*, pp 111-114. Eds D H Copp and R V Talmage. New York: Excerpta Medica.

17   Sternberger LA, Hardy Jr PH, Cucuclis JJ and Meyer HG (1970) The unlabelled antibody enzyme method of immunohistochemistry: Preparation and properties of soluble antigen-antibody complex (horseradish-peroxidase–antihorseradish-peroxidase) and its use on identification of spirochetes. *Journal of Histochemistry and Cytochemistry* **18** 315-333.

18   Urist M R (1961) Calcium and phosphorus in the blood and skeleton of the elasmobranchii. *Endocrinology* **69** 778-801.

The Comparative Endocrinology of Calcium Regulation
Eds C Dacke, J Danks, I Caple & G Flik, pp 91-96
Journal of Endocrinology Ltd, Bristol (1996)

Comparative renal tubular localization of calbindin-28k and the plasma membrane calcium pump of various species

T L Pannabecker, C A Smith, E J Braun and R H Wasserman

Cornell University, Ithaca, NY, USA and University of Arizona, Tucson, AZ, USA

Introduction

The transcellular transport of calcium across the intestinal epithelia and renal distal tubules is a multi-step process, involving the entrance of calcium across the luminal-facing brush border membrane, transport through the cell and uphill extrusion across the basolateral membrane (8, 23). The mechanism of transport at each of the three steps of transepithelial flux has been extensively studied, particularly in the rat and chicken intestine. In these species, intestinal epithelial transport is a vitamin D-dependent process and at least three vitamin D-dependent events have been identified. These include the increased permeability of the apical membrane (8), the induction of synthesis of calbindin-D_{28k} ($CaBP_{28k}$) in the chicken and calbindin-D_{9k} ($CaBP_{9k}$) in the rat (21), and the increased net synthesis of the basolateral plasma membrane calcium pump (PMCA) (24). In rat kidney, these same factors appear to be involved in the reabsorption of calcium from distal tubules shown to contain both calbindin-D_{28k} (9, 10, 17) and a relatively high concentration of the PMCA (2). The immunocytochemical co-localization of CaBP and relatively high concentrations of PMCA in various epithelial tissues (2, 22, 23) has suggested a functional relationship between these two proteins and the calcium-transporting capacity of epithelial tissues (4, 6, 19). However, we had earlier observed that this paradigm did not seem to hold for the avian kidney wherein, as in the rat kidney, $CaBP_{28k}$ was positioned in the distal tubules but PMCA was present in relatively high concentration in the proximal tubule. The regional separation of $CaBP_{28k}$ and PMCA in the avian kidney prompted the present study in which the comparative localization of immunoreactive $CaBP_{28k}$ and immunoreactive PMCA in the kidneys of a mammal (rat), bird (chicken), reptile (gecko) and an amphibian (toad) was determined.

Methods

Immunoreactivity of PMCA and $CaBP_{28k}$ by Western analysis

Kidneys, removed from rat, chicken, gecko and toad following decapitation, were processed by previously described procedures (22, 24). The tissue was polytron homogenized in buffer and, following homogenization, was centrifuged. The supernate was subsequently used for analysis of $CaBP_{28k}$ immunoreactivity, and the washed pellet was assessed for PMCA immunoreactivity. The proteins in the supernate and in the pellet were separated by polyacrylamide gel electrophoresis, and the proteins in each gel transferred electrophoretically to nitrocellulose for immunoblotting. The anti-PMCA antibody, designated 5F10 (2), is a monoclonal antibody produced against the human red cell PMCA and was either a kind gift from J T Penniston, Mayo clinic, Rochester, MN, or purchased from Affinity Bioreagents, Neshanic Station, NJ. The anti-CaBP$_{28k}$ polyclonal antibody was produced in rabbits. The appropriate peroxidase-conjugated second antibodies were from BioRad, Melville, NY.

Immunocytochemistry

Kidney tissue from the four species were fixed in 10% formalin in neutral saline for at least two days prior to embedding in paraffin. PMCA and CaBP$_{28k}$ were localized in tissue sections (approximately 5 µm thick) by immunoperoxidase staining with the streptavidin/biotin technique (antibody dilution: 1:100-1:2000). Co-localization in rat and chicken kidneys was investigated with double antibody labelling (antibody dilution, 1:25), using rhodamine- and fluorescein-conjugated second antibodies for the detection of PMCA and CaBP$_{28k}$ respectively, by confocal microscopy (5).

Results

Immunoblot analysis

Polyclonal antibodies raised against the purified chicken CaBP$_{28k}$ recognized a CaBP$_{28k}$ epitope in the supernatant fraction of the kidney homogenates of each of the four species examined. Immunoreactive proteins from the kidneys of the four species migrated essentially as a single band with an apparent molecular weight of about 28 kDa. Immunoreactive CaBP$_{28k}$ was previously detected in the kidneys of rat (10, 17), chicken (17), amphibian (12, 15) and reptile (12, 13). In some species, both the CaBP$_{28k}$ and calbindin-D$_{9k}$ (CaBP$_{9k}$) are present in renal tissue (14, 16, 18) and, in one species, the fruit bat, only CaBP$_{9k}$ is expressed (11).

Immunoreactive PMCA was also detected in the renal tissue of each of the four species. Multiple bands were usually observed, with apparent molecular weights that ranged from 127 to 145 kDa. The immunoreactive protein bands of human erythrocytes showed an apparent molecular weight of 142 kDa for the major band, and 147 and 135 kDa for the two minor bands. Multiple bands might reflect the presence of different isoforms of PMCA, or degradation and/or aggregation products.

Immunocytochemical localization of PMCA

For rat (Fig. 1A) and toad kidneys, the most prominent PMCA immuno-staining was in the distal tubules, with infrequent staining seen in proximal portions of the collecting duct of the toad. In the sections of rat kidney, the most intense staining clearly occurred along the basolateral domains of the cells of the distal tubules.

For the chicken (Fig. 1B) and gecko kidneys, dense staining was present in the proximal tubules, with considerably less staining in the distal segments. A diffuse pattern of staining was observed throughout the cytoplasm of the tubular cells of the chicken and the gecko, and moderate labeling of apparent endosomal membranes was observed with variable frequency. The brush border, which is much more prominent in proximal tubules than in distal tubules, showed little or no reaction product.

The glomeruli were devoid of immunoreactive product in all species. Tissue sections treated with non-immune ascites fluid were completely devoid of staining.

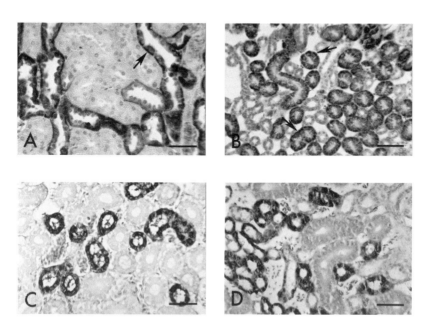

Fig. 1 Immunohistochemical localization of PMCA (A, B) and CaBP$_{28k}$ (C, D) in paraffin-embedded sections of kidneys from adult rat (A, C) and four week old chicken (B, D). The label for PMCA was heaviest along basolateral membranes of principal transporting cells (arrows) of rat distal tubules and chicken proximal tubules. The label for CaBP$_{28k}$ was observed only within cells of distal tubules in both rat and chicken. Anti-PMCA antibody diluted 1:2000 (rat) or 1:1000 (chicken). Anti-CaBP$_{28k}$ antibody diluted 1:500. The bar indicates 50 μm.

Immunocytochemical localization of CaBP$_{28k}$

Dense staining for CaBP$_{28k}$ was found primarily within the distal tubules of each of the four species, shown for the rat and chicken in Fig. 1C and 1D, respectively. Weaker staining was seen in the collecting duct of the gecko and toad kidney but no staining was seen in the collecting duct of the chicken or rat. At the cellular level, diffuse reaction products were apparent throughout the cytoplasm of the labeled cells, and no patterns of concentrated staining were seen along plasma membranes of the epithelial cells. As seen with the PMCA antibody, moderate labeling of apparent intracellular membranes was observed in distal nephrons of the chicken and gecko kidneys, and also infrequently in distal nephrons of the toad kidney.

The proximal tubules and glomeruli of all species did not react with the anti-CaBP$_{28k}$ antibody. Tissue sections treated with non-immune rabbit serum were completely devoid of staining.

Immunolocalization of CaBP$_{28k}$ and PMCA in rat and chicken kidney, using fluorescent double labeling and confocal microscopy

The confocal images revealed again the apparent exclusive localization of CaBP$_{28k}$ in the distal tubules of the rat and chicken. By this technique, the proximal tubules of the rat kidney did show some reactivity with the anti-PMCA antibody, but at a much lower intensity than in the distal tubules. For the chicken, the proximal tubules, as with the peroxidase method, showed a relatively high concentration of the PMCA epitope; immuno-reactive PMCA of distal tubules did become evident with this procedure, due perhaps to the apparent increased sensitivity of the fluorescent antibody procedure.

Discussion

The 5F10 anti-PMCA monoclonal antibody used in this study was developed against the PMCA of the human erythrocyte (2), and shown to cross-react with PMCA epitopes in other species and tissues. As examples, the reactivity of the 5F10 antibody with a PMCA epitope of rat renal distal tubules had been previously shown by Borke *et al.* (2). Immunoreactivity of the PMCA monoclonal antibody with chicken intestine (24) and oviduct (22) preceded the present observations on the avian kidney. The cross-reactivity of the 5F10 antibody with the PMCA of other non-mammalian vertebrates, such as the gecko and toad, has not been previously reported.

The major sites of renal localization of immunoreactive PMCA did appear to vary semi-quantitatively and qualitatively among the species examined. The most prominent immuno-staining of PMCA occurred in the distal tubules of the rat as well as the toad, whereas PMCA was most prominently detected in the proximal tubules of the chicken and gecko. CaBP$_{28k}$ was specifically localized in the distal tubular segment of all four species.

The lack of detailed physiological information on renal behavior in each of the four species precludes speculations on the overall relationship of tubular Ca^{2+} reabsorptive processes to the immuno-localization of PMCA and CaBP$_{28k}$. In the mammal, however, the distal tubules have generally been identified as the principal site of active Ca^{2+} reabsorption and the proximal tubules primarily the site of paracellular, diffusive processes (1, 7). PMCA and CaBP$_{28k}$ in calcium-transporting epithelia, including renal distal tubules, have been considered to operate cooperatively, the plasma membrane calcium pump actively extruding Ca^{2+} from the cell into the interstitial fluid and CaBP$_{28k}$ facilitating the movement of Ca^{2+} from the brush border region to the basolateral plasma membrane (4, 6, 23). The presence of CaBP$_{28k}$ and PMCA in the distal tubule of the toad does suggest that this might be an important site of active reabsorption of Ca^{2+} in this species.

The situation is less clear for the two other species, the chicken and the gecko, in which PMCA is present in high concentration in the proximal tubule, whereas CaBP$_{28k}$ is exclusively localized in the distal nephron, the site at which some active transport most likely does take place. The physiological consequence of the absence of $CaBP_{28k}$ in the PMCA-containing proximal tubules of the chicken and gecko is not known, but may indicate that another mechanism or system of transcellular Ca^{2+} transport is present and/or that there is a low transepithelial Ca^{2+}-transport rate in this segment. In any event, it appears that in this segment, PMCA functions independently of, or is less reliant upon the presence of, $CaBP_{28k}$.

Further investigations into the physiology of renal Ca^{2+} transport in non-mammalian vertebrates would certainly be of interest and possibly provide information pertinent to the findings in the present immunocytochemical study. Although not done here, consideration should be given to the possible contributions to renal tubular Ca^{2+} reabsorption of other Ca^{2+} extrusion processes, such as Na^+/Ca^{2+} exchange (1,7), as well as the role of the smaller calbindin $CaBP_{9k}$ in transepithelial Ca^{2+} transport. The direct stimulation of the plama membrane Ca^{2+} pump by $CaBP_{9k}$ has been reported (3, 20).

Finally, it was noted during the course of these studies that there is an apparent relationship between the site of the major localization of PMCA immunoreactivity and the form in which nitrogen is excreted in the various species studied here. The chicken and gecko primarily excrete uric acid and, in these species, PMCA is most prominent in the proximal renal tubules. The rat and toad primarily excrete urea and, in these species, PMCA is most prominent in the distal tubules. The possibility of more than a fortuitous relationship between these renal events might be considered.

References

1   Bindels RJM (1993) Calcium handling by the mammalian kidney. *Journal of Experimental Biology* **184** 89-104.

2   Borke JL, Caride A, Verma AK, Penniston JT & Kumar R (1989) Plasma membrane calcium pump and 28-kDa calcium binding protein in cells of rat kidney distal tubules. *American Journal of Physiology* **257** F842-F849.

3   Bouhtiauy I, Lajeunesse D, Christakos S & Brunette MG (1994) Two vitamin D3-dependent calcium binding proteins increase calcium reabsorption by different mechanisms II. Effect of CaBP 9K. *Kidney International* **45** 469-474.

4   Bronner F, Pansu D & Stein WD (1986) An analysis of intestinal calcium transport across the rat intestine. *American Journal of Physiology* **250** G561-G569.

5   de Talamoni NT, Smith CA, Wasserman RH, Beltramino C, Fullmer CS & Penniston JT (1993) Immunocytochemical localization of the plasma membrane calcium pump, calbindin-D_{28k}, and parvalbumin in Purkinje cells of avian and mammalian cerebellum. *Proceedings of the National Academy of Sciences of the USA* **90** 11949-11953.

6   Feher JJ, Fullmer CS & Wasserman RH (1992) The role of facilitated diffusion of calcium by calbindin in intestinal calcium absorption. *American Journal of Physiology* **262** C517-C526.

7   Friedman PA & Gesek FA (1993) Calcium transport in renal epithelial cells. *American Journal of Physiology* **264** F181-F198.

8   Fullmer CS (1992) Intestinal calcium absorption: calcium entry. *Journal of Nutrition* **122** 644-650.

9   Hermsdorf CL & Bronner F (1975) Vitamin D-dependent calcium-binding protein from rat kidney. *Biochimica et Biophysica Acta* **379** 553-561.

10   Intrator S, Elion J, Thomasset M & Brehier A (1985) Purification, immunological and biochemical characterization of rat 28 kDa cholecalcin (cholecalciferol-induced calcium-binding proteins). Identity between renal and cerebellar cholecalcins. *Biochemical Journal* **231** 89-95.

11   Opperman LA & Ross FP (1990) The adult fruit bat (Rousettus Aegyptiacus) expresses only calbindin-D9k (vitamin D-dependent calcium-binding protein) in its kidney. *Comparative Biochemistry & Physiology* **97B** 295-299.

12   Parmentier M, Ghysens M, Rypens F, Lawson DE, Pasteels JL & Pochet R (1987) Calbindin in vertebrate classes; immunohistochemical localization and Western blot analysis. *General & Comparative Endocrinology* **65** 399-407.

13   Rhoten WB, Lubit B & Christakos S (1984) Avian and mammalian vitamin D-dependent calcium binding protein in reptilian nephron. *General & Comparative Endocrinology* **55** 96-103.

14   Rhoten WB, Bruns ME & Christakos S (1985) Presence and localization of two vitamin D-dependent calcium binding proteins in kidneys of higher vertebrates. *Endocrinology* **117** 674-683.

15   Rhoten WB, Gona O & Christakos S (1986) Calcium-binding protein (28 000 Mr calbindin-D28k) in kidneys of the bullfrog Rana catesbeiana during metamorphosis. *Anatomical Record* **216** 127-132.

16   Schreiner DS, Jande SS, Parkes CO, Lawson DE & Thomasset M (1983) Immunocytochemical demonstration of two vitamin D-dependent calcium-binding proteins in mammalian kidney. *Acta Anatomica (Basel)* **117** 1-14.

17   Taylor AN, McIntosh JE & Bourdeau JE (1982) Immunocytochemical localization of vitamin D-dependent calcium-binding protein in renal tubules of rabbit, rat and chick. *Kidney International* **21** 765-773.

18   Taylor AN, Gleason WA Jr & Lankford GL (1984) Renal distribution of the 10 000 dalton vitamin D-dependent calcium-binding protein in neonatal rats. *Progress in Clinical & Biological Research* **168** 199-204.

19   van Os CH (1987) Transcellular calcium transport in intestinal and renal epithelial cells. *Biochimica et Biophysica Acta* **906** 195-222.

20   Walters JRF (1988) Regulatory effects of Ca-binding proteins on the Ca-pump of rat intestinal basolateral membranes. *Progress in Clinical & Biological Research* **252** 237-242.

21   Wasserman RH (1992) The intestinal calbindins: Their function, gene expression and modulation in genetic disease. *Extra- and Intracellular Calcium and Phosphate Regulation*, pp 43-70. Eds F Bronner & M Peterlik. Boca Raton, FL: CRC Press Inc.

22   Wasserman RH, Smith CA, Smith CM, Brindak ME, Fullmer CS, Krook L, Penniston JT & Kumar R (1991) Immunohistochemical localization of a calcium pump and calbindin-D_{28k} in the oviduct of the laying hen. *Histochemistry* **96** 413-418.

23   Wasserman RH, Chandler JS, Meyer SA, Smith CA, Brindak ME, Fullmer CS, Penniston JT & Kumar R (1992) Intestinal calcium transport and calcium extrusion processes at the basolateral membrane. *Journal of Nutrition* **122** 662-671.

24   Wasserman RH, Smith CS, Brindak ME, de Talamoni N, Fullmer CS, Penniston JT & Kumar R (1992) Vitamin D and mineral deficiencies increase the plasma membrane calcium pump of chicken intestine. *Gastroenterology* **102** 886-894.

The Comparative Endocrinology of Calcium Regulation
Eds C Dacke, J Danks, I Caple & G Flik, pp 97-102
Journal of Endocrinology Ltd, Bristol (1996)

Molecular cloning of complementary DNAs encoding the *Xenopus laevis* (Daudin) PTH/PTHrP receptor

C Bergwitz, P Klein[1] and H Jüppner

Endocrine Unit, Department of Medicine, Massachusetts General Hospital
and Harvard Medical School, Boston, MA, USA and [1]Department of
Molecular and Cellular Biology, Harvard University, Cambridge, MA, USA

Parathyroid hormone (PTH) and PTH-related peptide (PTHrP) act through a common G protein-coupled receptor which contains seven membrane-spanning helices, and activates at least two second messenger systems, protein kinase A and phospholipase C (1, 4). Complementary DNAs encoding the PTH/PTHrP receptor were isolated so far from five mammalian species, human (16, 17), mouse (7), rat (1, 14), pig (18) and opossum (4).

PTH, the major endocrine regulator of calcium and phosphate homeostasis in mammals, promotes in the kidney the 1α-hydroxylation of 25-vitamin D_3, the tubular reabsorption of calcium and secretion of phosphate, and it stimulates osteoblast function and thus bone re-modeling (11). PTHrP was discovered as the major cause of the humoral hypercalcemia of malignancy syndrome (2). It is, however, found in the circulation of healthy individuals, and its production by a variety of fetal and adult tissues suggests an important, but until now poorly understood, biological role as an autocrine/paracrine factor (2). This hypothesis is further supported by the finding that PTHrP and the PTH/PTHrP receptor are both expressed early during embryogenesis (7). Furthermore, mice that lack both alleles of either the PTHrP or the PTH/PTHrP receptor gene die shortly after birth or during mid-pregnancy respectively (6, 9), and display severe abnormalities of endochondral bone formation.

The endocrine regulation of calcium and phosphate homeostasis, which is very similar among tetrapods, differs significantly in fish (13). These vertebrates rely on a constant source of calcium in the environmental water and regulate its uptake through the gills by several hormones, i.e. prolactin, somatolactin, and stanniocalcin (3, 5). The presence of PTH-like substances in the pituitary gland and in the corpuscles of Stannius was previously reported, but their importance in the control of calcium homeostasis in fish remains unclear (3). Parathyroid glands are not found in fishes and first appeared in 'non-gill-breathing' vertebrates that moved onto land when bone and kidneys became important for the calcium homeostasis. Unlike in mammals, the pituitary gland, however, remains important for the control of calcium homeostasis in some amphibians, i.e. the more primitive urodeles and larval anurans, which may explain why parathyroidectomy is relatively well tolerated in this class of vertebrates (19). Therefore, amphibia have an intermediate position in the evolution of the vertebrate calcium metabolism. In *Xenopus laevis*, the four parathyroid glands develop from the third and fourth visceral pouch, and can already be seen as separate anatomic entities at the Nieuwkoop stage 23 (NF St 23) (1 days, 0.45 h) and NF St 25 (1 days, 3.5 h), respectively. Parathyroid epithelium first appears at NF St 43 (3 days, 15 h), the glands detach from the visceral pouches by NF St 59 (45 days), and are thought to assume their function during metamorphosis (>50 days) (12). In several anuran and urodelan species, including *Xenopus laevis*, the parathyroid glands involute during winter and regenerate during spring (15). While the biological action of PTH is similar in amphibians and in higher vertebrate species, nothing is known about the existence and/or function of PTHrP, although it is likely to be present in this vertebrate class.

To isolate cDNAs encoding the PTH/PTHrP receptor of the aquatic toad *Xenopus laevis* (Anura: Pipidae), we used a unidirectional λUniZAP(XR) cDNA library (Stratagene, La Jolla, CA), constructed from adult *Xenopus laevis* kidney mRNA, as a template to first amplify a DNA fragment of about 400 bp by polymerase chain reaction (PCR) (Fig. 1).

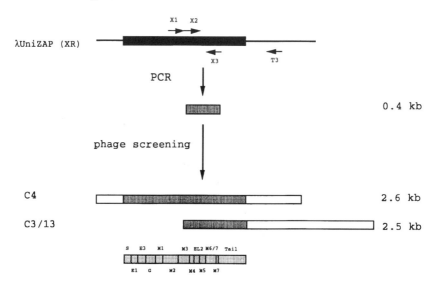

Fig. 1 Cloning strategy: the diagram at the bottom indicates the approximate location of the mammalian exons (8) in comparison to the *Xenopus laevis* PTH/PTHrP receptor clones.

Primer X1, which was based on the highly conserved, third membrane-spanning helix of all known mammalian species, and a primer specific for the vector's T3 promotor, were used for the first round of PCR-amplification. Re-amplification was then performed with nested primers that were based on other conserved DNA sequences (X2 and X3). The resulting 400 bp fragment was subcloned and identified as the *Xenopus* PTH/PTHrP receptor by nucleotide sequencing. This fragment was then used to screen the same cDNA library under high stringency conditions. Out of ≈400 000 plaque forming units (PFUs) three positive clones were isolated. Two partial clones, C3 and C13 (both ≈2.5 kb in length) are identical and encode the carboxyl-terminal portion of the frog PTH/PTHrP-receptor. Clone C4 (≈2.6 kb in length) encodes a full-length *Xenopus laevis* PTH/PTHrP-receptor containing 539 amino acids. This receptor shares high homology with the human, rat and opossum receptors, and contains most conserved residues that are likely to be of functional importance. The transmembrane domains, the intracellular loops and the third extracellular loop are overall better conserved than the extracellular N-terminus and the intracellular tail. Interestingly, the frog receptor homolog does not contain a region corresponding to the mammalian exon E2, which can be deleted without loss of function and which has the highest variability among the mammalian receptors (10)(Fig. 2). Using reverse-transcription PCR amplification of that region, we showed that clone C4 represents the major mRNA species in nine different tissues of adult *Xenopus laevis* .

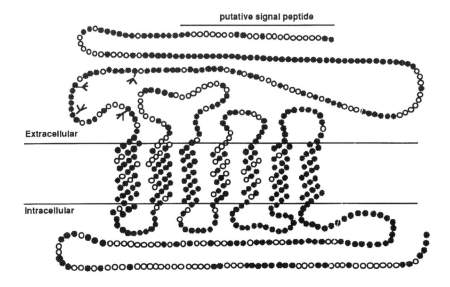

Fig. 2 Schematic representation of the PTH/PTHrP receptor: amino acids that are conserved in all known PTH/PTHrP receptor species including *Xenopus laevis* (stippled circles) or that are deleted in the *Xenopus laevis* receptor homolog (filled circles) are indicated.

The frog PTH/PTHrP receptor was transiently expressed in African green monkey kidney (COS-7) cells. Preliminary experiments showed high-affinity binding of radiolabeled [Nle^{8,18}, Tyr^{34}]bPTH(1-34)NH$_2$ (PTH(1-34)) and [Tyr^{36}]hPTHrP(1-36)NH$_2$ (PTHrP(1-36)) respectively. Competitive inhibition of radioligand binding was observed with PTH(1-34) and PTHrP(1-36), and to a lesser extent with PTH(3-34),

PTHrP(7-34), and PTH(7-34). Both PTH(1-34) and PTHrP(1-36) activated the adenylate cyclase in transfected COS-7 cells with equal efficacy, while PTHrP(1-36) appeared to be slightly more potent in activating phospholipase C. PTH(39-84) and unrelated peptides showed no measurable biological activity in all three assays.

In comparison to clone C4, the clones C3/13 contain several polymorphisms in the overlapping coding region and in the overlapping 3' non-coding region. Furthermore, the 3' non-coding region of clone C3/C13 is 900 bp longer than that of clone C4. A Southern blot of genomic DNA from *Xenopus laevis*, which was digested with several infrequently cutting restriction enzymes, was probed with this 900 bp 3' fragment and the full-length cDNA clone. Some of the DNA species hybridized with either probe indicating that the unique 3' non-coding region belongs to one of two frog PTH/PTHrP receptor genes. The presence of at least two non-allelic PTH/PTHrP receptor genes in *Xenopus laevis* was, furthermore, confirmed by re-probing these Southern blots with small DNA fragments that correspond to single mammalian exons (E3 and M2).

The lack of a proper outgroup makes an estimation of the divergence time of the two genes difficult. However, the relatively small number of nucleotide differences found in the overlapping coding region suggests that the duplication of the PTH/PTHrP receptor gene occurred when a genome duplication led to the speciation of *Xenopus laevis* approximately 30 million years ago (20). The presence of two PTH/PTHrP receptors genes may, therefore, be limited to the 'Laevis-group'.

In summary, we isolated a cDNA clone, C4, which encodes the full-length *Xenopus laevis* PTH/PTHrP receptor. This receptor is highly homologous to the known mammalian receptors, but misses a region in the amino-terminal, extracellular domain which corresponds to the mammalian exon E2. It binds mammalian PTH and PTHrP with similar affinity, and activates two second messenger systems, adenylate cyclase and phospholipase C. The partial clones, C3 and C13, are similar but not identical to the full-length C4 clone, and most likely represent the transcript of a second, non-allelic PTH/PTHrP receptor gene.

References

1   Abou-Samra AB, Jüppner H, Force T, Freeman MW, Kong XF, Schipani E, Urena P, Richards J, Bonventre JV, Potts JT Jr, Kronenberg H M & Segre GV (1992) Expression cloning of a common receptor for parathyroid hormone and parathyroid hormone-related peptide from rat osteoblast-like cells: a single receptor stimulates intracellular accumulation of both cAMP and inositol triphosphates and increases intracellular free calcium. *Proceedings of the National Academy of Sciences of the USA* **89** 2732-2736.

2   Broadus AE & Stewart AF (1994) Parathyroid hormone-related protein: structure, processing, and physiological actions. *The Parathyroids. Basic and Clinical Concepts*, pp 259-294. Eds JP Bilzikian, MA Levine & R Marcus. New York: Raven Press.

3   Hirano T (1989) The corpuscles of Stannius. *Regulation of Calcium and Phosphate*, vol 3, pp 139-169. Eds PKT Pang & MP Schreibman. San Diego: Academic Press.

4   Jüppner H, Abou-Samra AB, Freeman MW, Kong XF, Schipani S, Richards J, Kolakowski LF Jr, Hock J, Potts JT Jr, Kronenberg HM & Segre GV (1991) A G protein-linked receptor for parathyroid hormone and parathyroid hormone-related peptide. *Science* **254** 1024-1026.

5   Kaneko T & Hirano T (1993) Role of prolactin and somatolactin in calcium regulation in fish. *Journal of Experimental Biology* **184** 31-45.

6   Karaplis AC, Luz A, Glowacki J, Bronson R, Tybulewicz V, Kronenberg HM & Mulligan RC (1994) Lethal skeletal dysplasia from targeted disruption of the parathyroid hormone-related peptide gene. *Genes and Development* **8** 277-289.

7   Karperien M, van Dijk TB, Hoeijmakers T, Cremers F, Abou-Samra AB, Boonstra J, de Laat SW & Defize LHK (1994) Expression pattern of parathyroid hormone/parathyroid hormone related peptide receptor mRNA in mouse postimplantation embryos indicates involvement in multiple developmental processes. *Mechanisms of Development* **47** 29-42.

8   Kong XF, Schipani E, Lanske B, Joun H, Karperien M, Defize LHK, Jüppner H, Potts JT, Segre GV, Kronenberg HM & Abou-Samra AB (1994) The rat, mouse and human genes encoding the receptor for parathyroid hormone and parathyroid hormone-related peptide are highly homologous. *Biochemical and Biophysical Research Communications* **200** 1290-1299.

9   Lanske B, Karaplis AC, Luz A, Mulligan RC & Kronenberg HM (1994) Characterization of mice homozygous for the PTH/PTHrP receptor gene null mutation. *Journal of Bone and Mineral Research* **9** (Suppl 1) S121.

10 Lee C, Gardella TJ, Abou-Samra AB, Nussbaum SR, Segre GV, Potts JT Jr, Kronenberg H M & Jüppner H (1994) Role of the extracellular regions of the parathyroid hormone (PTH)/PTH-related peptide receptor in hormone binding. *Endocrinology* **135** 1488-1495.

11 Murray TM, Rao LG & Rizzoli RE (1994) Interactions of parathyroid hormone, parathyroid hormone-related protein, and their fragments with conventional and nonconventional receptor sites. *The Parathyroids,* pp 185-211. Eds JP Bilezikian, MA Levine & R Marcus. New York: Raven Press.

12 Nieuwkoop PD & Faber J (1967) *Normal Table of Xenopus Laevis (Daudin).* Amsterdam: North-Holland.

13 Pang PTK & Pang RK (1989) Hormones and calcium regulation in vertebrates: an evolutionary and overall consideration. *Regulation of Calcium and Phosphate*, vol 3, pp 343-352. Eds PKT Pang & MP Schreibman. San Diego: Academic Press.

14 Pausova Z, Bourdon J, Clayton D, Mattei M-G, Seldin TMF, Janicic N, Riviere M, Szpirer J, Levan G, Szpirer C, Goltzman D & Hendy GN (1994) Cloning of a parathyroid hormone/parathyroid hormone-related peptide receptor (PTHR) cDNA from a rat osteosarcoma (UMR106) cell line: chromosomal assignment of the gene in the human, mouse, and rat genomes. *Genomics* **20** 20-26.

15 Robertson D (1977) The annual pattern of plasma calcium in the frog and the seasonal effect of ultimobranchialectomy and parathyroidectomy. *Journal of Comparative Endocrinology* **33** 336-343.

16 Schipani E, Karga H, Karaplis AC, Potts JT Jr, Kronenberg HM, Segre GV, Abou-Samra AB & Jüppner H (1993) Identical complementary deoxyribonucleic acids encode a human renal and bone parathyroid hormone (PTH)/PTH-related peptide receptor. *Endocrinology* **132** 2157-2165.

17   Schneider H, Feyen JH, Seuwen K & Movva NR (1993) Cloning and functional expression of a human parathyroid hormone receptor. *European Journal of Pharmacology* **246** 149-155.

18   Smith DP, Hsiung HM & Zang X (1994) Structure and functional expression of a complementary DNA for porcine parathyroid receptor. *GenBank Accession# U18315.*

19   Stiffler DF (1993) Amphibian calcium metabolism. *Journal of Experimental Biology* **184** 47-61.

20   Tymowska J (1991) Polyploidity and cytogenetic variation in frogs of the genus Xenopus. *Amphibian Cytogenetics and Evolution*, pp 259-298. Eds DM Green & SK Sessions. San Diego: Academic Press.

The Comparative Endocrinology of Calcium Regulation
Eds C Dacke, J Danks, I Caple & G Flik, pp 103-105
Journal of Endocrinology Ltd, Bristol (1996)

Parathyroid hormone-related protein in the saccus vasculosus of fishes

P M Ingleton and J A Danks[1]

Institutes of Endocrinology and Cancer Studies, The Medical School, Beech Hill Road, Sheffield S10 2RX, UK and [1]St Vincent's Institute of Medical Research, 41 Victoria Parade, Fitzroy, Victoria 3065, Australia

The saccus vasculosus (SV) is a prominent organ lying beneath the brain and posterior to the pituitary in some teleost and elasmobranch fishes, made obvious by the red blood cells filling the dense network of capillaries within the tissue. The lumen of the SV is continuous with that of the third ventricle of the mid-brain and hence is filled with cerebrospinal fluid (CSF). Separating the capillary space from the lumen is a layer of epithelial cells, often only one cell deep, which have multiple protrusions into the luminal fluid forming coronets. The coronets are expanded, closed ends of cilia of the epithelial cells creating an enormously increased surface area for exchange between the luminal fluid and the blood capillaries via the epithelial cells. There is also an extensive network of nerves within the saccus so that it is a neurohaemal organ both from its internal anatomy and through the position of the coronet cells lying between the neurotransmitter-containing CSF and capillaries (1).

Although the SV is a prominent neurohaemal organ its specific function has not been closely studied and no bioactive factors have been isolated from or detected in the coronet cells. Early studies proposed that it was concerned with providing a store of glycogen for brain tissue (9) but it has also been considered to be a tissue for monitoring CSF ionic composition (4), particularly mono and divalent cations, because the cells possess ion-specific ATPase enzymes. There is also evidence that the tubular system of the coronets is open to the SV lumen allowing uptake of larger molecules from the CSF (3).

We have been studying the distribution of parathyroid hormone-related protein (PTHrP) in fishes using antibodies to human PTHrP and have noted that coronet cells of several species reacted with these antibodies.

Materials and methods

Brain and pituitaries of sea bream, Spanish mackerel, flounder and dogfish were fixed in sublimated Bouin-Hollande (6), dehydrated, cleared and wax embedded. Immunoreactive PTHrP (irPTHrP) was demonstrated in fixed tissue by the technique of Sternberger (8) as described in Danks *et al.* (2), using primary rabbit antisera to human PTHrP 1-16, 50-69 and pre-pro- -13 -- +2; normal rabbit serum or primary antiserum blocked with PTHrP peptide was used as a negative control. Sea bream SVs were also fixed in glutaraldehyde and paraformaldehyde (5) and some fragments post-fixed in osmium tetroxide before embedding in e-mix. Tissue not post fixed was used for immunodetection of PTHrP. Sections were mounted on nickel grids, etched with sodium ethoxide, blocked with normal goat serum and reacted with anti-human PTHrP 1-16 for 2 h at room temperature; specific binding was detected with gold-labelled second antibody.

Sea bream SVs were also extracted in phosphate buffered saline (pH 7.6) and extracted proteins separated on 15% gels by SDS/PAGE (7). After separation gels were stained with Coomassie blue or the proteins transferred by electro-transfer to nitrocellulose paper for Western blotting. SVs of sea bream were also incubated in Krebs-Ringer bicarbonate and the medium collected for SDS/PAGE and Western blotting after 24 h and a further 48 h. The same primary antisera used for immunohistochemistry were also used to detect irPTHrP proteins in SV extracts and culture media.

Results
Coronet cells of the saccus of all the fish species examined, including the elasmobranch dogfish, stained with the PTHrP antisera, but not all the cells contained irPTHrP. Ultrastructurally the coronet cells of sea bream resembled those described for other fishes with extensive smooth endoplasmic reticulum (SER) in cell and coronet. Specific immunogold reactions detected irPTHrP in association with SER membranes in both coronets and cell bodies.

SDS/PAGE separated several proteins in sea bream SV extracts, with fewer at lower intensity in the incubation media. The proteins in greatest concentrations had molecular weights of approximately 14.3 and 15 kDa. These two proteins in media and SV extracts, both reacted with antisera to human PTHrP 1-16, 50-69 and −13-+2. The molecular weights of these proteins suggest they are similar to human PTHrP 1-84.

Conclusions
Both teleost and elasmobranch SV coronet cells contain irPTHrP. The fact that irPTHrP is detectable in tissue extracts and incubation media of sea bream SVs by antisera to human PTHrP 1-16, 50-69 and −13-+2 suggests it is similar in amino composition to human PTHrP in the N-terminus, mid-molecule and pre-pro regions. Some of the coronet epithelial cells appear to be the source of this irPTHrP. The presence of irPTHrP in the incubation medium of SVs suggests that it may be synthesised or processed by SV cells. The physico-chemical characteristics of irPTHrP from other species need to be examined and physiological studies are needed to determine the possible function of irPTHrP in the SV of fishes. Nevertheless this is the first report of a potentially bioactive factor in the saccus vasculosus of fishes.

Acknowledgements
Some of this work was supported by EC grant AQ2.458; JAD is a JD Wright Fellow of the National Health and Medical Research Council of Australia; PMI receives financial support from the Yorkshire Cancer Research Campaign.

References

1   Altner H & Zimmerman H (1972) The saccus vasculosus. *Structure and Function of Nervous Tissue*, vol 5, pp 293-328. Ed GH Bourne. New York & London: Academic Press.

2   Danks JA, Devlin AJ, Ho PMW, Diefenbach-Jagger H, Power DM, Canario A, Martin TJ & Ingleton PM (1993) Parathyroid hormone-related protein is a factor in normal fish pituitary. *General Comparative Endocrinology* **92** 201-212.

3   Follenius E (1982) Relationships between the tubular system in the globules of the coronet cells of the saccus vasculosus and the cerebrospinal fluid in *Gasterosteus aculeatus* form *leiurus* (Teleostei). *Cell and Tissue Research* **224** 105-115.

4   Jansen WF, Flight WFG & Zandbergen MA (1981) Fine structural localisation of adenosine triphosphatase activities in the saccus vasculosus of the rainbow trout. *Cell and Tissue Research* **219** 267-279.

5   Karnovsky MJ (1965) A formaldehyde-glutaraldehyde fixative of high osmolality for use in electron microscopy. *Journal of Cellular Biology* **27** 137A.

6   Kraicer J, Herlant M & Duclos P (1967) Changes in adenohypophyseal cytology and nucleic acid content in the rat 32 days after bilateral adrenalectomy and chronic injection of cortisol. *Canadian Journal of Physiology* **45** 947-956.

7   Laemmli UK (1970) Cleavage of structural proteins during assembly of the head of bacteriophage T4. *Nature* **227** 680-685.

8   Sternberger LA (1974) *Immunocytochemistry.* Englewood-Cliffs, New Jersey: Prentice-Hall.

9   Sundaraj BI & Prasad MRN (1963) The histophysiology of the saccus vasculosus of *Notopterus chilata* (Teleostei). *Quarterly Journal of Microscopy Science* **104** 465-469.

The Comparative Endocrinology of Calcium Regulation
Eds C Dacke, J Danks, I Caple & G Flik, pp 107-109
Journal of Endocrinology Ltd, Bristol (1996)

Parathyroid hormone-related protein in the corpuscles of Stannius

J A Danks, R J Balment[1] and P M Ingleton[2]

St Vincent's Institute of Medical Research, Fitzroy, Victoria 3065,
Australia; [1]Department of Physiological Sciences, University of Manchester
M13 9PT, UK and [2]Institutes of Endocrinology and Cancer Research,
Medical School, Sheffield, S10 2RX, UK

The corpuscle of Stannius (CS) are organs found only in ray-finned fish, both holostean and teleostean. The glands lie in the tissue of the kidney, the numbers varying with species, and embryologically appear to originate from epithelial cells associated with the nephric ducts (4, 5). Two cell types have been identified in several species based on ultrastructural features, with type 1 cells, which are the most numerous and have large secretory granules and type 2 cells which have only small secretory granules and are much fewer in number (8, 9, 16). Induction of hypocalcaemia has been identified as the principal function of the corpuscles and is associated with the type 1 cells, whilst type 2 cells may be concerned with sodium and potassium ion conservation in media of low ionic strength (8, 15). The hypocalcaemic factors have been isolated and identified from several teleosts (1, 13, 14); stanniocalcin is a glycopeptide varying in molecular weight from 39 to 60 kDa; a second hypocalcaemic factor, teleocalcin, is much smaller, 3 kDa, and appears to influence branchial Ca^{2+} and Mg^{2+}-dependent phosphatase activity. A third factor, parathyrin, has been isolated from eel (*Anguilla anguilla*) CS and shown to cause mobilisation of bone in the rat but stimulate gill Ca^{2+} efflux in the eel (10, 11). This factor resembles mammalian parathyroid hormone.

We have been examining tissues, including corpuscles of Stannius, from teleosts for the presence of immunoreactve parathyroid hormone-related protein (irPTHrP) following the identification of this factor in serum and pituitary of the sea bream (2).

Materials and methods

CS tissue from holosteans, *Amia calva* and *Lepisosteus* (gift of Dr John Youson) and teleosts eel (*Anguilla anguilla*), rainbow trout (*Oncorhynchus mykiss*), chum salmon (*Oncorhynchus keta*) (gift of Dr G Wagner) and seawater (SW)- and freshwater (FW)-adapted flounder (*Platichthys flesus*) were fixed in sublimated Bouin-Hollande (7), dehydrated, cleared and wax embedded. irPTHrP was demonstrated in fixed tissue by the technique of Sternberger (12) as described in Danks *et al.* (2), using primary rabbit antiserum to human PTHrP(1-16); normal rabbit serum or primary antiserum blocked with PTHrP peptide were used as negative controls. The primary anti-human PTHrP(1-16) has been extensively characterised and does not react with PTH in mammalian systems (3). Flounder CS were also fixed in glutaraldehyde/paraform-aldehyde (6) and some fragments post-fixed in osmium tetroxide before embedding. Tissue not post-fixed was used for immunodetection of PTHrP. Sections were mounted on nickel grids, etched with sodium ethoxide, blocked with gelatin and normal swine serum and reacted with anti-human PTHrP(1-16) for 2 h at room temperature; specific binding was detected with gold-labelled goat anti-rabbit antibody.

Results

There was no immune reaction in any of the cells of either of the holostean CS; similarly neither salmonid, trout nor chum salmon, CS cells reacted with anti PTHrP(1-16) antiserum. Kidney tubules of *Amia* and *Lepisosteus* reacted weakly with the antiserum. It was therefore surprising to find that cells in both FW and SW flounder CS reacted with the anti-PTHrP antiserum. The cells were situated principally in the periphery of the pseudofollicles, mainly showing staining in the pole of the cell adjacent to the capillary. There was no obvious difference between the number and distribution of irPTHrP containing cells in FW- and SW-adapted fish.

Ultrastructurally the cells of the flounder CS resembled those of other teleosts with two cell types distinguishable by the size of the secretory granules. The cells contained abundant rough endoplasmic reticulum (RER) and secretory granules, the type 1 cells having numerous large granules and type 2 cells only a few small granules. Ultrastructural immunohistochemistry with antiserum to human PTHrP(1-16) revealed that in both FW- and SW-adapted corpuscles granules of a sub-population of type 1 cells contained irPTHrP.

Conclusions

irPTHrP can be detected in CS cells of the euryhaline flounder (*Platichthys flesus*) with an antiserum to human PTHrP(1-16). Other species of teleost and two holostean do not appear to have the same factor in CS cells.

Comments

The irPTHrP factor in the flounder may have greater amino acid homology with human PTHrP than the other species, or the absence of reaction may indeed indicate the absence of any similar peptide. The embryological origin of the cells of the CS may determine their ability to synthesise PTHrP. In the flounder, kidney tubule cells and ureter epithelial cells react strongly with antiserum to human PTHrP(1-16) whilst in other species in this study, reaction in kidney tubules was weak, suggesting a relationship between them and the CS cells. It is also possible that there is a similarity between the 'primitive' salmonids and holosteans whilst the flounder may be considered more 'advanced' in these feature (8).

References

1   Butkus A, Roche PJ, Fernley RT, Haralambidis J, Penschow JD, Ryan GB, Tahair JF, Tregear GW & Coghlan JP (1987) Purification and cloning of a corpuscles of Stannius protein from *Anguilla australis*. *Molecular and Cellular Endocrinolgy* **54** 123-133.

2 Danks JA, Devlin AJ, Ho PMW, Diefenbach-Jagger H, Power DM, Canario A, Martin TJ & Ingleton PM (1993) Parathyroid hormone-related protein is a factor in normal fish pituitary. *General and Comparative Endocrinology* **92** 201-212.

3 Danks JA, Ebeling PR, Hayman J, Chou ST, Moseley JM, Dunlop J, Kemp BE & Martin TJ (1989) Parathyroid hormone-related protein: immunochemical localisation in cancers and in normal skin. *Journal of Bone and Mineral Research* **4** 273-278.

4 de Smet D (1962) Considerations on the Stannius corpuscles and interrenal tissues of bony fishes, especially based on researches into *Amia. Acta Zoologica (Stockholm)* **43** 201-219.

5 Kaneko T, Hasegawa S & Hirano T (1992) Embryonic origin and development of the corpuscles of Stannius in chum salmon (*Oncorhynchus keta*). *Cell and Tissue Research* **268** 65-70.

6 Karnovsky MJ (1965) A formaldehyde-glutaraldehyde fixative of high osmolality for use in electron microscopy. *Journal of Cellular Biology* 27 137A

7 Kraicer J, Herlant M & Duclos P (1967) Changes in adenohypophyseal cytology and nucleic acid content in the rat 32 days after bilateral adrenalectomy and the chronic injection of cortisol. *Canadian Journal of Pharmacology and Physiology* **45** 947-956.

8 Krishnamurthy Vg & Bern HA (1969) Correlative histologic study of the corpuscles of Stannius and the juxtaglomerular cells of teleost fishes. *General and Comparative Endocrinology* **13** 313-335.

9 Meats M, Ingleton PM, Chester Jones I, Garland HO & Kenyon CJ (1978) Fine structure of the corpuscles of Stannius of the trout *Salmo gairdneri*: structural changes in response to increased environmental salinity and calcium ions. *General and Comparative Endocrinology* **36** 451-461.

10 Milet C, Hillyard CJ, Martelly E, Girgis S, MacIntyre I & Lopez E (1980) Similitudes structurales entre l'hormone hypocalcemiante des corpuscles de Stannius (PCS) de l'Anguille (*anguilla anguilla* L) et l'hormone parathyroidienne mammalienne. *Comptes Rendus de L'Academie des Sciences Paris* **291** 977-980.

11 Milet C, Martelly E & Lopez E (1989) Partial purification of parathyrin from the corpuscles of Stannius (PCS) of the eel (*Anguilla anguilla*, L). *General and Comparative Endocrinology* **76** 83-94.

12 Sternberger LA (1974) *Immunocytochemistry.* Englewood-Cliffs, NJ: Prentice-Hall.

13 Verbost PM, Butkus A, Willems P & Wendelaar Bonga SE (1993) Indication for two bioactive principles in the corpuscles of Stannius. *Journal of Experimental Biology* **177** 243-252.

14 Wagner GF, Hamphong M, Park CM & Copp DH (1986) Purification, characterization and bioassay of teleocalcin, a glycoprotein from salmon corpuscles of Stannius. *General and Comparative Endocrinology* **63** 481-491.

15 Wendelaar Bonga SE, Greven JAA & Veenhuis M (1976) The relationship between the ionic composition of the environment and the secretory activity of the endocrine cell types of Stannius corpuscles in the teleost *Gasterosteus aculeatus. Cell and Tissue Research* **175** 297-312.

16 Wendelaar Bonga SE, van der Meij JCA & Pang PKT (1980) Evidence for two secretory cell types in the Stannius bodies of the teleosts *Fundulus heteroclitus* and *Carassius auratus. Cell and Tissue Research* **212** 295-306.

Part Two

Dinosaurs and birds

The Comparative Endocrinology of Calcium Regulation
Eds C Dacke, J Danks, I Caple & G Flik, pp 113-121
Journal of Endocrinology Ltd, Bristol (1996)

Avian bone turnover and the role of bone cells

C V Gay

Departments of Poultry Science and Biochemistry & Molecular Biology
The Pennsylvania State University, University Park, PA 16802, USA

Introduction

The goal of this overview is to tie some new evidence and ideas into some of the older concepts, rather than to provide an exhaustive review. The areas discussed will focus mainly on the actions of parathyroid hormone (PTH), calcitonin and estrogen on bone cells. Effects of vitamin D on avian bone have been recently described (37). The roles of cytokines are beginning to be understood, particularly from mammalian cell studies (36, 61).

The avian skeleton is distinguished from the mammalian in several important ways. Notably, in avians, bone formation is rapid, beginning at around day 6 of embryogenesis when osteoblasts first appear on the surfaces of the cartilage anlage (3). By the time of hatching, on day 21, well-mineralized bones have developed. Rapid growth continues, and by sexual maturity, ~18 weeks for female laying hens, a unique type of bone, medullary bone, develops as outgrowths from the endosteal surface into the marrow cavities of ribs, vertebrae, femurs and tibias (48). Medullary bone, which serves as a labile reservoir for eggshell calcium, is the most physiologically responsive of all types of bone in vertebrates (13).

In discussing the role of bone cells in avian bone turnover, several types of bone should be considered: cortical, medullary and trabecular. One model system in use is the growing chick tibia from which osteoclasts can be isolated from the endosteum (29) and osteoblasts from the periosteal surface (26). Medullary bone is a valuable source of osteoclasts since large, relatively pure populations of cells can be readily obtained (13). Little attention has been given to trabecular bone in avian species, perhaps because medullary bone is so much more responsive.

Bone cell surfaces are believed to be completely covered with cells, either quiescent bone lining cells or osteoblasts and few or many osteoclasts, depending on the demand for resorption (30). The surface layer is believed to be contiguous with osteocytes buried within the matrix.

The role of the lining cells is considered to be twofold: first, to serve as a barrier that separates bone fluid space from the general extracellular fluid, and second, to develop into osteoblasts upon receiving correct signals. Osteoblasts synthesize a calcifiable matrix and also serve in defining the bone fluid space. Bone matrix consists of type I collagen and key non-collagenous proteins such as osteocalcin, osteonectin, fibronectin and other adhesive proteins (53). Osteoblasts secrete and are regulated by numerous regulatory factors (43, 49). Correctly formed matrix calcifies when Ca^{++} and HPO_3^- are present in appropriate concentrations. In medullary bone, osteoblasts produce less collagen and it is poorly organized (13). Medullary bone is also unique in that substantial amounts of proteoglycan, mainly keratan sulfate, are present (12, 21). These differences between medullary and cortical bone in collagen and proteogylcan content apparently allow for greater deposition of mineral, at the expense of strength, which is not needed (13).

Calcium flux and cells on bone surfaces: role of PTH

It is widely believed, although not unequivocally proven, that bone surfaces are lined with a continuous layer of cells. Definitive corroboration of this concept has been provided by an ultrastructural study that used lanthanum as an electron-opaque tracer (50). Near-term fetal rat calvaria were treated with lanthanum nitrate during fixation. Lanthanum had access only to very early osteogenic sites in which initial mineralization occurred in matrix vesicles. Following this stage of development, osteoblasts became tightly adherent and lanthanum only partially penetrated the intercellular spaces between osteoblasts; mineralizing osteoid was devoid of the tracer. It can be concluded from this study that, after initial mineral deposition, mineralization of osteoid proceeds in a compartment that is distinct from the general extracellular fluid.

A study on cultured osteoblasts also shows that osteoblasts adhere tightly to one another (28). A confluent layer of primary rat pup calvarial osteoblasts was placed in a chamber designed to monitor transmembrane flow rates. The rate was found to be low and comparable to that for sheets of cultured capillary endothelial cells. Transmembrane flux was diminished further, by 90%, when the cells were treated with PTH.

A number of studies show that osteoblasts respond to PTH by changing shape. Cultured chick periosteal osteoblasts increased in area within one minute of PTH treatment (32) as did cultured rat pup calvarial osteoblasts (28). A spread in cell area would cause osteoblasts arranged in a sheet to push tightly together and reduce transmembrane fluid movement. Not all studies are in agreement with these observations, however. For example, osteoblasts cultured to confluence have been observed to become stellate when treated with PTH (35).

In order to understand the means by which Ca^{++} is delivered to mineralizing surfaces, we have begun to evaluate the mechanisms of Ca^{++} uptake and release by osteoblasts. Plasma membrane Ca-ATPase (pmATPase) was found by ultrastructural histochemistry of chick metaphysis to be present along the apico-lateral sides of the cell and absent along the cell surface in contact with osteoid (2). Vesicles prepared from plasma membrane of cultured chick periosteal osteoblasts translocated Ca^{++} at a rate of 9.9 ± 2.3 nmol/mg protein/min (24). This transport rate is similar in magnitude to rates found in many cells, for example red blood cells, and it is substantially less than the rate found in intestine and kidney, tissues which utilize Ca-ATPase to absorb or reabsorb massive quantities of Ca^{++} (9). The direction of Ca^{++} transport by osteoblast plasma membrane vesicles was outward (24), as has been reported for all other tissues studied (22). On the basis of location as well as direction and magnitude of pumping, it appears that the pmCa-ATPase in osteoblasts is not involved in delivery of Ca^{++} to sites of mineralization.

More likely, the role of pmCa-ATPase in osteoblasts is to restore cytosolic Ca^{++} to basal levels following hormone treatment. A calcium-specific fluorescent dye, Calcium Green C_{18} (Molecular Probes Inc., Eugene OR) was loaded into the plasma membrane of chick periosteal osteoblasts (32). The calcium chelating head group was facing out so that as Ca^{++} emerged from the cell it would be trapped and cause the dye to fluoresce. Stimulation with a bolus of PTH (approximate concentration 10^{-8} M) resulted in peripheral cell fluorescence in ten seconds; fluorescence intensity peaked in 2-3 min and was spent in 9-10 min. Fluorescence intensity was substantially reduced by sodium vanadate, quercetin and trifluoperazine, indicating that the translocated Ca^{++} was largely dependent on Ca-ATPase. Thapsigarin, which blocks re-entry of Ca^{++} into intracellular stores, enhanced the duration of the response, indicating that maintenance of cytosolic Ca^{++} levels involved both intracellular stores and plasma membrane efflux.

Osteoclast activation and the role of PTH

Bone modeling and remodeling requires the coupled interaction of both osteoblasts and osteoclasts. According to the Rodan and Martin hypothesis (46), osteoblasts move aside to allow osteoclast podosomes to attach to bone surfaces. It has become well established that more osteoclasts associate with bone surfaces during physiological and pathological bone resorption (60). However, in medullary bone the numbers of osteoclasts do not fluctuate but appear to oscillate between active and inactive states (58, 59). We estimate that for growing chick tibia ~50% of the endosteal surface is occupied by osteoclasts. It is likely that the cells of the tightly adherent osteoblast-bone lining cell layer discussed in the previous section move apart in discrete locales to allow attachment of osteoclasts. The process would likely be regulated by specific local cytokines.

The main task of an osteoclast is to destroy calcified bone and it does so by secreting both proteolytic enzymes and acid into the resorption lacuna (14). The source of the hydrogen ions is from carbonic anhydrase activity, as we and others have shown (23). Inhibition of carbonic anhydrase blocks bone resorption (14, 23). For example, when there was an infusion of acetazolamide into rats with humoral hypercalcemia of malignancy, plasma Ca^{++} was restored to normal, whereas healthy control rats developed hypocalcemia (10). Osteoclasts are regulated by several systemic hormones and numerous local regulatory factors, both physiologic and pathologic (14, 51). One of the earliest recognized controlling substances is PTH.

Many studies indicate that PTH acts on osteoclasts indirectly and is mediated by factors secreted by neighboring cells, i.e. osteoblasts or stromal cells of the marrow (43, 49). While osteoblasts are widely held as the pivotal cell in regulating bone metabolism, at least one avian and one mammalian study have identified neighboring cells, and excluded osteoblasts, as the mediator (47, 54).

The evidence that PTH also interacts directly with osteoclasts is sparse. Four studies have shown specific binding of PTH to osteoclasts through immunocytochemistry of rat bone sections (44), or application of radiolabeled (16, 55) or biotinylated-PTH (1) to avian osteoclasts. The immunocytochemical study was appropriately controlled and the binding studies met the criteria for specific binding, namely saturability, high-affinity binding and competition by the native ligand. As a test of function, 95% pure preparations of isolated osteoclasts have been neutralized with 20 mM NH_4OH, then stained with acridine orange to detect hydrogen ions, and the capacity of the cells to re-acidify has been monitored in the presence and absence of PTH (25, 29, 34). These studies consistently found a stimulation of osteoclast acidification by PTH. The cells examined were attached to coverslips and were microns to millimeters away from the few contaminating cells present. Therefore, it seems unlikely that a small number of distally situated osteoblasts or other cells could control hundreds of osteoclasts. This data, combined with the evidence that PTH binds specifically to osteoclasts, suggests that osteoclasts are directly contolled by PTH, in addition to being influenced indirectly.

Figure 1 presents a hypothetical role for the direct action of PTH on osteoclasts. As discussed in the previous section, PTH causes osteoblasts to extrude Ca^{++} (32). Because of the close proximity of osteoblasts and osteoclasts *in vivo* it is possible that the concentration of Ca^{++} adjacent to osteoclasts rises substantially when osteoblasts have been stimulated with PTH. Recent studies from Zaidi's group indicate that both avian and mammalian osteoclasts possess calcium receptors and that receptor occupancy down-regulates osteoclast activity, although avian osteoclasts required time in culture before the response was detectable (6). We suggest that the inhibitory effect of Ca^{++}, derived from PTH-stimulated osteoblasts, is overridden by the ability of osteoclasts to respond directly to PTH.

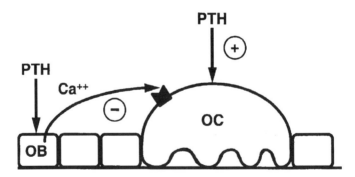

Fig. 1 Diagram depicting that Ca^{++} is released by osteoblasts (OB) when exposed to PTH as previously described (32) and that osteoclasts (OC) are inhibited when their calcium receptors are occupied (6). It is postulated that the inhibitory effect of calcium is overcome by PTH directly stimulating the osteoclast.

As understanding of parathyroid-hormone-related peptide (PTHrP) has developed, it has become evident that PTH and PTHrP utilize the same receptor due to high homology of the amino terminus of each peptide (57). The carboxy terminus of PTHrP has been shown to inactivate both isolated rat (19) and chicken osteoclasts (20). Interaction of PTHrP with the PTH receptor of bone cells is an emerging and exciting area of investigation.

Estrogen action on bone cells

Studies on the interaction of estrogen, specifically 17β-estradiol (E_2), with bone cells present some interesting inconsistencies. For osteoblasts, the evidence for the presence of the classic nuclear/cytoplasmic estrogen receptor (ER) is quite disparate. High-affinity saturable binding site assays indicate both low levels of receptor, at ~200 sites per nucleus (31), and high levels, at 1600 sites per nucleus (18). Specific binding of ^{3}H-estradiol to discrete sites along the endosteum of E_2-induced male Japanese quail has been detected (56). The sites are presumed to be where E_2-induced medullary bone will form. The most convincing evidence is the demonstration of the receptor itself, e.g. by immunostaining. In Japanese quail medullary bone, in both females and E_2-induced males, osteoblasts, pre-osteoblasts and bone lining cells have been found positive for ER by immunostaining and binding of fluorescently tagged estrogen (38-40). However, in sections of pig, guinea pig and human bone from donors at the correct stage of estrous or menses, Braidman and colleagues (8) found a distinct absense of ER in osteoblasts by immunostaining. Both avian (39, 40) and mammalian (8) osteocytes have strikingly high levels of ER by immunostaining. Since osteocytes derive from osteoblasts, it seems likely that under certain unique conditions osteoblasts would also express ER as, for example, at the onset of medullary bone formation.

The nuclear/cytoplasmic E_2 receptor has been demonstrated in avian medullary bone osteoclasts by the high-affinity nuclear binding assay (5662 sites per nucleus), by cDNA probing for mRNA and by Western blotting (42). E_2 has also been found to reduce acidification quickly (25). This rapid response to E_2 led to the proposal that a plasma membrane E_2 receptor exists in osteoclasts (11). An E_2-bovine serum albumin complex conjugated to fluorescein (E_2-BSA-Fl) was detected on the surface of osteoclasts using confocal microscopy (11). Rapid responses to the complexed E_2 were noted, namely changes in cell shape and reduced acidification. To achieve a maximal effect, 2 μM E_2-BSA-FL was needed; 1 μM E_2 was adequate to block binding of the complex. Additional studies on estrogen effects on bone cells are described in a recent review (41).

Action of calcitonin on avian osteoclasts

While some studies have reported an absence of calcitonin receptors on avian osteoclasts, a number of others have indicated the converse. At the whole animal level, the antibody to calcitonin infused into 19-day chick embryos caused mild hypercalcemia (5). Calcitonin injected into laying hens when an eggshell was not being calcified caused hypocalcemia; however, during eggshell formation calcitonin was ineffective (33). This observation suggests that medullary bone osteoclast inhibition by calcitonin can be overridden, a process that may be unique to medullary bone. In three-week old chicks, osteoclast morphology was affected by infused calcitonin in the predicted manner, namely the cells rounded up and their ruffled borders became attenuated (4).

Isolated avian osteoclasts have been found to respond to calcitonin in several studies. These include a reduction in acidification (25, 29), a cyclic AMP response (17, 45), cell retraction (17), and inhibition of resorption pit formation (15). Focal adhesion kinase expression was suppressed by calcitonin in osteoclasts from 18 day chick embryos, as well as in human osteoclasts (7). Osteoclasts on cultured fragments of medullary bone lost their ruffled borders and actin filament orientation under the influence of calcitonin (52).

Weak, but measurable, binding of biotinylated calcitonin to osteoclasts has been shown (27). Two differences were noted in comparison to the biotinylated PTH study. For calcitonin, lower levels of fluorescence were observed and more rapid surface clearance, believed to be endocytosis, occurred.

While these studies indicate that calcitonin influences osteoclasts under a wide variety of conditions, there are differences between avian and mammalian osteoclast responses to calcitonin, particularly with respect to medullary bone. However, it should be noted that avian cortical bone osteoclasts seem to be quite mammalian-like. Determining what the differences are at the signal pathway level will be of considerable value.

Summary

This overview has focused on avian bone cell studies and where possible compared the results to mammalian studies. The effects of PTH on osteoblasts have also been considered, as has the evidence that PTH stimulates osteoblasts to release Ca^{++} on the marrow-facing side. The direct stimulation of osteoclasts by PTH has been considered and its potential role in overriding the down-regulation of osteoclasts by Ca^{++} has been postulated. The presence of nuclear estrogen receptors in osteoblasts, osteocytes and osteoclasts was reviewed and the discovery of plasma membrane estrogen receptors in osteoclasts discussed. The effects of calcitonin on avian osteoclasts have been reviewed.

Acknowledgements

Editorial assistance by Virginia R Gilman and Joseph P Stains and support by NIH Grants No. DE04345 and DE09459 is gratefully acknowledged.

References

1   Agarwala N & Gay C (1992) Specific binding of parathyroid hormone to living osteoclasts. *Journal of Bone and Mineral Research* **7** 531-539.

2   Akisaka T, Yamamoto T & Gay C (1988) Ultracytochemical investigation of calcium-activated adenosine triphosphatase (Ca^{++}-ATPase) in chick tibia. *Journal of Bone and Mineral Research* **3** 19-25.

3   Anderson H (1973) Calcium-accumulating vesicles in the intercellular matrix of bone. *Hard Tissue Growth, Repair and Remineralization (Ciba Foundation Symposium 11)*, pp 213-246. Amsterdam: Elsevier.

4   Anderson R, Schraer H & Gay C (1982) Ultrastructural immunocytochemical localization of carbonic anhydrase in normal and calcitonin-treated chick osteoclasts. *Anatomical Record* **204** 9-20.

5   Baimbridge K & Taylor T (1980) Role of calcitonin in calcium homeostasis in the chick embryo. *Journal of Endocrinology* **85** 171-185.

6   Bascal Z, Alam A, Zaidi M & Dacke C (1994) Effect of raised extracellular calcium on cell spread area in quail medullary bone osteoclasts. *Experimental Physiology* **79** 15-24.

7   Berry V, Rathod H, Pulman L & Datta H (1994) Immunofluorescent evidence for the abundance of focal adhesion kinase in the human and avian osteoclasts and its down regulation by calcitonin. *Journal of Endocrinology* **141** R11-R15.

8   Braidman I, Davenport L, Carter D, Selby P, Mawer E & Freemont A (1995) Preliminary *in situ* identification of estrogen target cells in bone. *Journal of Bone and Mineral Research* **10** 74-80.

9   Bronner F (1991) Calcium transport across epithelia. *International Review of Cytology* **131** 169-212.

10   Brown G, Morris C, Mitnick M & Insogna K (1990) Treatment of humoral hypercalcemia of malignancy in rats with inhibitors of carbonic anhydrase inhibitors. *Journal of Bone and Mineral Research* **5** 1037-1041.

11   Brubaker K & Gay C (1994) Specific binding of estrogen to osteoclast surfaces. *Biochemical and Biophysical Research Communications* **200** 899-907.

12   Candlish J & Holt F (1971) The proteoglycans of fowl cortical and medullary bone. *Comparative Biochemistry and Physiology* **40B** 283-293.

13   Dacke C, Arkle S, Cook D, Wormstone I, Jones S, Zaidi M & Bascal Z (1993) Medullary bone and avian calcium regulation. *Journal of Experimental Biology* **184** 63-84.

14   Delaissé J & Vaes G (1992) Mechanism of mineral solubilization and matrix degradation in osteoclastic bone resorption. *Biology and Physiology of the Osteoclast*, pp 289-314. Eds BR Rifkin & CV Gay. Boca Raton, FL: CRC Press Inc.

15   de Vernejoul M, Horowitz M, Demignon J, Neff L & Baron R (1988) Bone resorption by isolated chick osteoclasts in culture is stimulated by murine spleen cell supernatant fluids (osteoclast-activating factor) and inhibited by calcitonin and prostaglandin E_2. *Journal of Bone and Mineral Research* **3** 69-80.

16   Duong L, Grasser W, DeHaven P & Sato M (1990) Parathyroid hormone receptors identified on avian and rat osteoclasts. *Journal of Bone and Mineral Research* **5** S203.

17   Eliam M, Baslé M, Bouizar Z, Bielakoff J, Moukhtar M & de Vernejoul M (1988) Influence of blood calcium on calcitonin receptors in isolated chick osteoclasts. *Journal of Endocrinology* **119** 243-248.

18   Eriksen E, Colvard D, Berg N, Graham M, Mann K, Spelsberg T & Riggs B (1988) Evidence of estrogen receptors in normal human osteoblast-like cells. *Science* **241** 84-86.

19   Fenton A, Kemp B, Kent G, Moseley J, Zheng M, Rowe D, Britto J, Martin T & Nicholson G (1991) A carboxyl-terminal peptide from the parathyroid hormone-related peptide inhibits bone resorption by osteoclasts. *Endocrinology* **129** 1762-1768.

20   Fenton A, Martin T & Nicholson G (1994) Carboxyterminal parathyroid hormone-related protein inhibits bone resorption by isolated chicken osteoclasts. *Journal of Bone and Mineral Research* **9** 515-519.

21 Fisher L & Schraer H (1982) Keratan sulfate proteoglycan isolated from the estrogen-induced medullary bone in Japanese quail. *Comparative Biochemistry and Physiology* **72B** 227-232.

22 Garrahan P & Rega A (1990) Plasma membrane calcium pump. *Intracellular Calcium Regulation*, pp 271-303. Eds F Bronner. New York: Wiley-Liss.

23 Gay C (1992) Osteoclast ultrastructure and enzyme histochemistry: functional implications. *Biology and Physiology of the Osteoclast*, pp 129-150. Eds BR Rifkin & CV Gay. Boca Raton, FL: CRC Press Inc.

24 Gay C & Lloyd Q (1995) Characterization of calcium efflux by osteoblasts derived from long bone periosteum. *Comparative Biochemistry and Physiology* **111A** 257-261.

25 Gay C, Kief N & Bekker P (1993) Effect of estrogen on acidification in osteoclasts. *Biochemical and Biophysical Research Communications* **192** 1251-1259.

26 Gay C, Lloyd Q & Gilman V (1994) Characteristics and culture of osteoblasts dervived from avian long bone. *In Vitro Cellular Developmental Biology* **30A** 379-383.

27 Hall M, Kief N, Gilman V & Gay C (1994) Surface binding and clearance of calcitonin by avian osteoclasts. *Comparative Biochemistry and Physiology* **108A** 59-63.

28 Hillsley M & Frangos J (1996) Osteoblast hydraulic conductivity is regulated by calcitonin and parathyroid hormone. *Journal of Bone and Mineral Research* **11** 114-124.

29 Hunter S, Schraer H & Gay C (1988) Characterization of isolated and cultured chick osteoclasts: the effects of acetazolamide, calcitonin and parathyroid hormone on acid production. *Journal of Bone and Mineral Research* **3** 297-303.

30 Jones S, Boyde A, Ali N & Maconnachie E (1985) A review of bone cell and substrate interaction. An illustration of the role of scanning electron microscopy. *Scanning* **7** 5-24.

31 Komm B, Terpening C, Benz D, Graeme K, Gallegos A, Korc M, Greene G, O'Malley B & Haussler M (1988) Estrogen binding, receptor mMNA, and biologic response in osteoblast-like osteosarcoma cells. *Science* **241** 81-84.

32 Lloyd Q, Kuhn M & Gay C (1995) Characterization of calcium translocation across the plasma membrane of primary osteoblasts using a lipophilic calcium-sensitive fluorescent dye, calcium green C_{18}. *Journal of Biological Chemistry* **270** 22445-22451.

33 Luck M, Sommerville B & Scanes C (1980) The effect of eggshell calcification on the response of plasma calcium activity to parathyroid hormone and calcitonin in the domestic fowl (*Gallus domesticus*). *Comparative Biochemistry and Physiology* **65A** 151-154.

34 May L, Gilman V & Gay C (1993) Parathyroid hormone regulation of avian osteoclasts. *Avian Endocrinology*, pp 227-238. Ed PJ Sharp. Bristol: Journal of Endocrinology Ltd.

35 Miller S, Wolf A & Arnaud C (1976) Bone cells in culture: morphologic transformation by hormones. *Science* **192** 1340-1342.

36 Mundy G (1993) Cytokines and growth factors in the regulation of bone remodeling. *Journal of Bone and Mineral Research* **8** S505-510.

37 Norman A & Hurwitz S (1993) The role of the vitamin D endocrine system in avian bone biology. *Journal of Nutrition* **123** 310-316.

38 Ohashi T, Kusuhara S & Ishida K (1990) Histochemical identification of oestrogen target cells in the medullary bone of laying hens. *British Journal of Poultry Science* **31** 221-224.

39 Ohashi T, Kusuhara S, Ishida K (1991) Immunoelectron microscopic demonstration of estrogen receptors in osteogenic cells of Japanese quail. *Histochemistry* **96** 41-44.

40 Ohashi T, Kusuhara S & Ishida K (1991) Estrogen target cells during the early stage of medullary bone osteogenesis: immunohistochemical detection of estrogen receptors in osteogenic cells of estrogen-treated male Japanese quail. *Calcified Tissue International* **49** 124-127.

41   Oursler M, Landers J, Riggs B & Spelsberg T (1993) Oestrogen effects on osteoblasts and osteoclasts. *Annals of Medicine* **25** 361-371.

42   Oursler M, Osdoby P, Pyfferoen J, Riggs B & Spelsberg T (1991) Avian osteoclasts as estrogen target cells. *Proceedings of the National Academy of Science* **88** 6613-6617.

43   Puzas J & Ishibe M (1992) Osteoblast/osteoclast coupling. *Biology and Physiology of the Osteoclast* pp 337-356. Eds BR Rifkin and CV Gay. Boca Raton, FL: CRC Press Inc.

44   Rao L, Murray T & Heersche J (1983) Immunohistochemical demonstration of parathyroid hormone binding to specific cell types in fixed rat bone tissue. *Endocrinology* **113** 805-810.

45   Rifkin B, Auszmann J, Kleckner A, Vernillo A & Fine A (1988) Calcitonin stimulates cAMP accumulation in chicken osteoclasts. *Life Sciences* **42** 799-804.

46   Rodan G & Martin T (1981) Role of osteoblasts in hormonal control of bone resorption: a hypothesis. *Calcified Tissue International* **33** 349-351.

47   Rouleau M, Mitchell J & Goltzman D (1990) Characterization of the major parathyroid hormone target cell in the endosteal metaphysis of rat long bones. *Journal of Bone and Mineral Research* **5** 1043-1053.

48   Schraer H & Hunter S (1985) The development of medullary bone: a model for osteogenesis. *Comparative Biochemistry and Physiology* **82A** 13-17.

49   Simmons D & Grynpas M (1990) Mechanisms of bone formation *in vivo*. *Bone Volume I The Osteoblast and Osteocyte*, pp 193-302. Ed BK Hall. Caldwell, NJ: The Telford Press.

50   Soares A, Arana-Chavez V, Reid A & Katchburian E (1992) Lanthanum tracer and freeze-fracture studies suggest that compartmentalization of early bone matrix may be related to initial mineralization. *Journal of Anatomy* **181** 345-356.

51   Stern P & Lakatos P (1992) Effects of pharmacologic agents on osteoclasts. *Biology and Physiology of the Osteoclast* pp 357-395. Eds BR Rifkin & CV Gay. Boca Raton: CRC Press Inc.

52   Sugiyama T & Kusuhara S (1996) Chapter in this volume.

53   Termine J (1993) Bone matrix proteins and the mineralization process. *Primer on the Metabolic Bone Diseases and Disorders of Mineral Metabolism*, 2nd edn, pp 21-25. Ed MJ Favus. New York: Raven Press.

54   Teti A, Grano M, Colucci S & Zambonin Zallone A (1991) Osteoblastic control of osteoclast bone resorption in a serum-free co-culture system. Lack of effect of parathyroid hormone. *Journal of Endocrinological Investigation* **15** 63-68.

55   Teti A, Rizzoli R & Zambonin Zallone A (1991) Parathyroid hormone binding to cultured avian osteoclasts. *Biochemical and Biophysical Research Communications* **174** 1217-1222.

56   Turner R, Bell N & Gay C (1993) Evidence that estrogen binding sites are present in bone cells and mediate medullary bone formation in Japanese quail. *Poultry Science* **72** 728-740.

57   Urena P, Kong X, Juppner H, Kronenberg H, Potts J & Segre G (1993) Parathyroid hormone (PTH)/PTH-related peptide receptor messenger ribonucleic acids are widely distributed in rat tissues. *Endocrinology* **133** 617-623.

58   van de Velde J, Vermeiden J, Touw J & Veldhuijzen J (1984) Changes in activity of chicken medullary bone cell populations in relation to the egg-laying cycle. *Metabolic Bone Disease and Related Research* **5** 191-193.

59   Zambonin Zallone A & Mueller W (1969) Medullary bone of laying hens during calcium depletion and repletion. *Calcified Tissue Research* **4** 136-146.

60   Zheng M, Nicholson G, Wharton A & Papadimitriou J (1991) What's new in osteoclast ontogeny? *Pathology, Research and Practice* **187** 117-125.

61   Zheng M, Wood D & Papadimitriou J (1992) What's new in the role of cytokines in osteoblast proliferation and differentiation? *Pathology, Research and Practice* **188** 1104-1121.

The Comparative Endocrinology of Calcium Regulation
Eds C Dacke, J Danks, I Caple & G Flik, pp 123-130
Journal of Endocrinology Ltd, Bristol (1996)

Ca^{2+}-sensing receptors in avian osteoclasts

M Zaidi[1], Z A Bascal[2], O A Adebanjo[1], S Arkle[2] and C G Dacke[2]

[1]Department of Medicine and Institute on Aging, University of
Pennsylvania, Philadelphia, PA 19104, USA and [2]Division of
Pharmacology, University of Portsmouth, Portsmouth, UK

Introduction

It is now widely accepted that the bone resorbing cell, the osteoclast, is sensitive to changes in the ambient Ca^{2+} concentration (8, 14). Such changes are known to occur locally as a consequence of bone mineral dissolution during resorption (12). In the isolated rat osteoclast, an increase in the extracellular Ca^{2+} concentration has been found to trigger an acute elevation in the cytosolic Ca^{2+} concentration. This is followed by a marked retraction of the osteoclast margin and a loss of the cell's secretory and bone resorptive activities. These observations have formed the basis of our initial suggestion that changes in the extracellular Ca^{2+} concentration are monitored by a unique sensor or 'receptor' for Ca^{2+} (9, 16).

Here, we summarize our investigations whereby we have used the Japanese quail, an oestrogen-sensitive bird, to study Ca^{2+} receptor expression in the avian osteoclast. Medullary bone of the Japanese quail undergoes unrestricted remodelling during egg-laying (7). It thus provides an accessible source of osteoclasts which, in contrast to the cortical and trabecular bone of adult mammalian species, such as the rat, are relatively easily isolated and purified. The ease of isolation of quail osteoclasts is because of the low mineral content of quail bones and the richly cellular nature of the medullary tissue. We found osteoclasts harvested from medullary bone during egg-laying did not express the Ca^{2+} receptor (5). Nevertheless, further studies showed that their sensitivity to extracellular Ca^{2+} was promptly restored when the cells were cultured *in vitro* away from the microenvironment of the long bone (4, 6). It has thus become evident that Ca^{2+} receptor expression is at least, in part, determined by the level of required bone resorption. These studies in the avian osteoclast have also strengthened our hypothesis for a primary role for this molecule in the regulation of bone remodelling.

Methods

Osteoclasts were isolated from medullary bone of female Japanese quail, *Coturnix coturnix japonica*. Inbred quail hens aged 3 to 10 months were kept on an 18 h light/6 h darkness cycle and were fed on a quail egg-laying diet (SDS). Following an overnight fast with access to water *ad libitum*, those having an egg in their shell gland were sacrificed by cervical dislocation. Osteoclasts were mechanically disaggregated from their femora and tibiae into pre-warmed (37 °C) N-2-hydroxyethyl piperazine-N-2-ethane sulfonic acid- (Hepes-) buffered Medium 199 (ICN Flow UK Ltd, Middlesex, UK). The cell suspension (1 ml per well) was dispersed onto 35 mm plastic Petri dishes or glass coverslips which were incubated in a humidified chamber at 37 °C for 60 to 120 min to allow for the cells to sediment on and attach to the substrate. The wells were then washed with Medium 199 to remove non-adherent contaminating cells and incubated for a further 20 to 60 min to allow for cell spreading.

Osteoclast spread area was assessed by a morphometric procedure described earlier by Zaidi *et al.* (17). Briefly, osteoclasts were visualized under a phase contrast microscope (Nikon, London, UK) that was linked through a charge coupled device (CCD) camera to a time-lapse video recorder. The composite video signal was fed into a 256 gray level imaging system (Sight Systems, Newbury, Berks, UK). Digitized gray images were captured into the computer memory at two-minute intervals followed by digitization of the cell boundary. Each cell image tracing was retrieved sequentially and was analyzed using a software package programmed to measure area within each tracing by a mathematical procedure (17).

Cytosolic [Ca²⁺] was measured by a fura-2- or indo-1-based ratiometric method. An inverted phase-contrast microscope fitted with an epifluorescence attachment (Diaphot, Nikon UK Ltd, Telford, UK) was used. For fura-2-based measurements (10), coverslips containing dye-loaded osteoclasts were exposed alternately to excitation wavelengths of 340 and 380 nm approximately every second by a micro-computer-driven wheel that contained band-pass interference filters. After passing through a dichroic mirror (400 nm), the transmitted fluorescence beam was filtered at 510 nm to be directed into a photomultiplier tube (PM28B, Thorn EMI, London, UK). The resulting single photon currents were converted to transistor-transistor logic pulses, counted and recorded in an IBM microcomputer. Finally, the ratio of emitted intensities due to excitation at 340 and 380 nm, F_{340}/F_{380}, was calculated and displayed. For indo-1 based measurements, we used a method described previously (14). Both dyes were calibrated by established protocols (3, 10, 14).

Results and discussion

Fig. 1 shows the effect of 5, 10 and 20 mM-Ca²⁺ on the spread area of isolated quail osteoclasts. Up to a 40 min exposure failed to reduce cell spread area at any one of the three concentrations tested (6). In contrast, freshly isolated rat osteoclasts showed a dramatic reduction in cell spread area when exposed to 20 mM extracellular Ca²⁺; there was a reduction to 50% of the original cell spread area within 20 min of the elevation of extracellular Ca²⁺ (1). This was followed by almost complete recovery of the cells even in the sustained presence of an elevated extracellular Ca²⁺ (Fig. 2). Both cell types were found to respond to ionomycin. Freshly isolated quail cells retracted to approximately about 80% of their original plan area, whilst rat osteoclasts showed a somewhat larger response magnitude (6, 15).

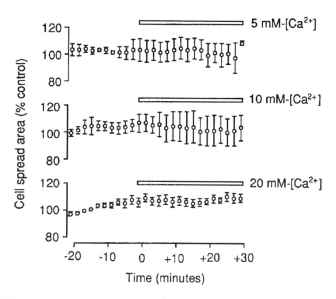

Fig. 1 Effects of elevated extracellular Ca^{2+} concentrations (5, 10 and 20 mM) on the spread area of freshly isolated medullary bone osteoclasts. Responses are shown as a percentage of control cell spread area (mean±standard error of the mean (SEM)). Reproduced with permission from (6).

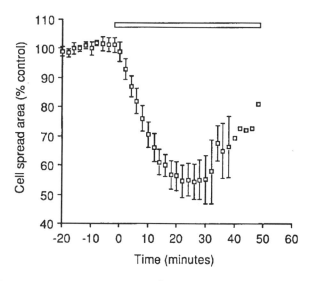

Fig. 2 Effect of an elevated extracellular Ca^{2+} concentration (20 mM, open bar) on the spread area of freshly isolated rat osteoclasts. Responses are shown as a percentage of control cell spread area (mean±SEM). Redrawn from (1).

125

The retained response of quail osteoclasts to the Ca^{2+} ionophore ionomycin clearly suggested that the machinery required for cell retraction in response to an elevated cytosolic Ca^{2+} concentration remained intact and that the defect was probably in the sensing of extracellular Ca^{2+}. We investigated this further in experiments whereby we measured changes in the cytosolic Ca^{2+} concentration in response to the extracellular application of Ca^{2+} as well as ionomycin. The results are summarized in Fig. 3, from which it is clear that the exposure of isolated quail cells to a high ambient Ca^{2+} concentration (5, 10 or 20 mM) failed to produce a cytosolic Ca^{2+} signal (5). In contrast, rat cells showed an almost doubling of their cytosolic Ca^{2+} levels with an increase in the extracellular Ca^{2+} concentration to 20 mM, as previously reported (14). However, as would be expected, both freshly isolated quail and rat osteoclasts showed changes in their cytosolic Ca^{2+} levels in response to the application of ionomycin. These findings confirmed our hypothesis that the defect was not in the responsiveness of these cells to changes in cytosolic Ca^{2+}, but instead arose from their inability to 'sense' a changed extracellular Ca^{2+} concentration.

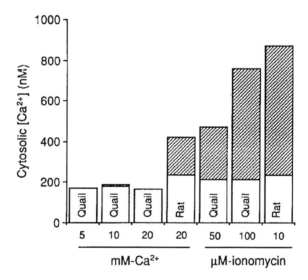

Fig. 3 Effect of various treatments with Ca^{2+} (mM) or ionomycin (μM) on cytosolic Ca^{2+} concentrations (nM) in indo-1-loaded freshly isolated quail or rat osteoclasts. The shaded bars represent post-treatment levels, whilst the open bars show basal levels.

Quite unexpectedly, we found that when quail osteoclasts were cultured for seven days, they began to demonstrate responses to changes in extracellular Ca^{2+} (Table 1). Figure 4 shows the response of cultured quail osteoclasts to an elevated extracellular Ca^{2+} concentration (20 mM). A dramatic cell retraction to values of around 50% of control values was noted, but unlike rat cells those cultured from the quail were found

Table 1 Differential sensitivities of the freshly isolated rat and quail osteoclasts, as well as quail osteoclasts cultured for 7 days, in terms of their ability to respond (+) or not to respond (-) to changes in the extracellular Ca^{2+} concentration (calcium, up to 20 mM) or application of the Ca^{2+} ionophore, ionomycin.

Species	Conditions	Calcium	Ionomycin
Rat	Fresh	+	+
Quail	Fresh	–	+
Quail	Cultured	+	+

not to recover spontaneously. When cytosolic Ca^{2+} levels were measured in monolayers of cultured cells, extracellular Ca^{2+} elevations produced prompt increases in cytosolic Ca^{2+}, in a manner dependent upon the cation concentration (Fig. 5). A complete recovery of osteoclasts in terms of their lost sensitivity to Ca^{2+}, we believe, represents the fact that we removed these cells from the microenvironment of the long bone. These findings provided the first clues that Ca^{2+} was being recognized in the osteoclast by a receptor-like entity which we termed the Ca^{2+} receptor. Expression of this putative receptor could be controlled, at least in the quail osteoclast, by the level of increased bone turnover required during egg-laying.

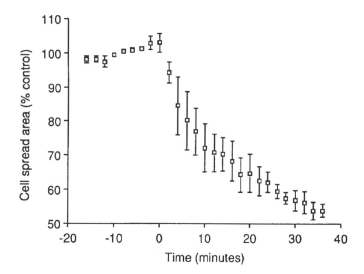

Fig. 4 Effect of an elevated extracellular Ca^{2+} concentration (20 mM) on the spread area of quail osteoclasts cultured for seven days. Responses are shown as a percentage of control cell spread area (mean±SEM). Reproduced with permission from (6).

127

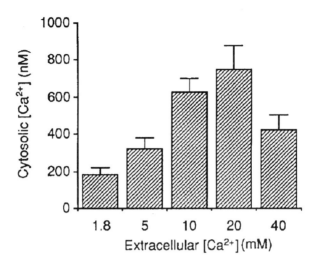

Fig. 5 Effect of elevated extracellular Ca^{2+} levels (1.8, 5, 10, 20 and 40 mM) on cytosolic Ca^{2+} concentrations (nM) in fura-2-loaded quail osteoclasts cultured for seven days. The levels are shown as mean±SEM ($n=4$ to 14 cells at each concentration). Redrawn from (4).

We have since attempted to characterize the activation properties of this 'receptor' using divalent cations other than Ca^{2+} using the rat osteoclast system. A classical pattern of drug-receptor interactions has emerged from these studies (10). For example, there is evidence of concentration-dependent activation as well as use-dependent inactivation. Our argument for the existence of a Ca^{2+} receptor has become more compelling by our demonstration that its occupancy triggers a distinct and experimentally separable intracellular Ca^{2+} release mechanism (10). We have examined whether a ryanodine receptor is involved in this process in a similar way to the mechanism described for voltage-sensing in striated muscle (18). Using a panel of isoform- and epitope-specific antibodies to the ryanodine receptor family, we have very recently demonstrated that a possibly novel form of the type II ryanodine receptor is localized to the osteoclast surface membrane (19).

Our studies have further shown that apart from Ca^{2+}, two other influences vital to the function of a resorbing osteoclast, namely its membrane potential and the ambient pH, can control Ca^{2+} receptor function. It is notable that an osteoclast is exposed to pH values as low as 4 units and can rest at either one of the two preferred membrane potentials of -15 and -70 mV, respectively (13). We found that either a low pH or a hyperpolarized membrane potential markedly enhanced Ca^{2+} receptor sensitivity (2, 11). Thus, it is possible that the Ca^{2+} receptor regulates osteoclast activity not only by sensing changes in the concentrations of Ca^{2+} and H^+, but also by monitoring simultaneous alterations in the cell's membrane potential.

References

1   Adebanjo OA, Pazianas M, Zaidi A, Shankar VS, Bascal ZA, Dacke CG, Huang CL-H & Zaidi M (1994) Quantitative studies on the effect of prostacyclin on freshly isolated rat osteoclasts in culture. *Journal of Endocrinology* **143** 375-381.

2   Adebanjo OA, Shankar VS, Pazianas M, Zaidi A, Huang CL-H & Zaidi M (1994) Modulation of the osteoclast Ca^{2+} sensing by extracellular protons. Possible linkage between Ca^{2+} sensing and extracellular acidification. *Biochemical and Biophysical Research Communications* **194** 742-747.

3   Alam ASMT, Bax CMR, Shankar VS, Bax BE, Bevis, PJR, Huang CL-H, Moonga BS, Pazianas M & Zaidi M (1993) Further studies on the mode of action of calcitonin on isolated rat osteoclasts. Pharmacological evidence for a second site mediating intracellular Ca^{2+} mobilization and cell retraction. *Journal of Endocrinology* **136** 7-15.

4   Arkle S, Wormstone IM, Bascal ZA & Dacke CG (1994) Estimation of intracellular calcium activity in confluent monolayers of primary cultures of quail medullary bone osteoclasts. *Experimental Physiology* **79** 975-982.

5   Bascal ZA, Moonga BS, Dacke CG & Zaidi M (1992) Osteoclasts from medullary bone of egg-laying Japanese quail do not express the putative calcium receptor. *Experimental Physiology* **77** 501-504.

6   Bascal ZA, Alam ASMT, Zaidi M & Dacke CG (1994) Effect of raised extracellular calcium on cell spread area in quail medullary bone osteoclasts. *Experimental Physiology* **79** 15-24.

7   Dacke CG (1989) Eicosanoids, steroids and miscellaneous hormones. In *Vertebrate Endocrinology: Fundamentals and Biomedical Implications*, vol 3, pp 171-210. Ed Pang PKT & Schreibman MP. New York: Academic Press.

8   Malgaroli A, Meldolesi J, Zambonin-Zallone A & Teti A (1989) Control of cytosolic free calcium in rat and chicken osteoclasts. *Journal of Biological Chemistry* **264** 14342-14347.

9   Moonga BS, Moss DW, Patchell A & Zaidi M (1990) Intracellular regulation of enzyme secretion from rat osteoclasts and evidence for a functional role in bone resorption. *Journal of Physiology* **429** 29-45.

10   Shankar VS, Bax CMR, Bax BE, Alam ASMT, Simon B, Pazianas M, Moonga BS, Huang CL-H & Zaidi M (1993) Activation of the Ca^{2+} receptor on the osteoclast by Ni^{2+} elicits cytosolic Ca^{2+} signals: evidence for receptor activation and inactivation, intracellular Ca^{2+} redistribution and divalent cation modulation. *Journal of Cellular Physiology* **155** 120-129.

11   Shankar VS, Huang CL-H, Adebanjo OA, Simon BJ, Alam ASMT, Moonga BS, Pazianas M, Scott RH & Zaidi M (1995) The effect of membrane potential on surface rat osteoclasts. *Journal of Cellular Physiology* **162** 1-8.

12   Silver IA, Murrils RJ & Etherington DJ (1988) Microelectrode studies on the acid micro-environment beneath adherent macrophages and osteoclasts. *Experimental Cell Research* **175** 266-276.

13   Sims SM & Dixon SJ (1989) Inwardly rectifying K^+ currents in osteoclasts. *American Journal of Physiology* **89** C1277-1282.

14   Zaidi M, Datta HK, Patchell A, Moonga BS & MacIntyre I (1989) Calcium-activated intracellular calcium elevation: a novel mechanism of osteoclast regulation. *Biochemical and Biophysical Research Communications* **163** 1461-1465.

15   Zaidi M, Datta HK, Moonga BS & MacIntyre I (1990) Evidence that the action of calcitonin on rat osteoclast is mediated by two G proteins acting via separate post-receptor pathways. *Journal of Endocrinology* **126** 473-481.

16   Zaidi M, Kerby J, Huang CL-H, Alam ASMT, Rathod H, Chambers TJ & Moonga BS (1991) Divalent cations mimic the inhibitory effects of extracellular ionized calcium on bone resorption by isolated rat osteoclasts: further evidence for a 'calcium receptor'. *Journal of Cellular Physiology* **149** 422-427.

17   Zaidi M, Alam ASMT, Shankar VS, Bax BE, Moonga BS, Bevis PJR, Pazianas M & Huang CL-H (1992) A quantitative description of components of *in vitro* morphometric change in the rat osteoclast model: relationships with cellular function. *European Biophysics Journal* **21** 349-355.

18   Zaidi M, Shankar VS, Alam ASMT, Moonga BS, Pazianas M & Huang CL-H (1992) Evidence that a ryanodine receptor triggers signal transduction in the osteoclast. *Biochemical and Biophysical Research Communications* **188** 1332-1336.

19   Zaidi M, Shankar VS, Tunwell R, Adebanjo OA, Mackrill J, Pazianas M, O'Connell D, Simon BJ, Rifkin BR, Venkitoraman AR, Huang CL-H & Lai FA (1995) A ryanodine receptor-like molecule expressed in the osteoclast plasma membrane functions in extracellular Ca^{2+} sensing. *Journal of Clinical Investigation* **96** 1582-1590.

The Comparative Endocrinology of Calcium Regulation
Eds C Dacke, J Danks, I Caple & G Flik, pp 131-135
Journal of Endocrinology Ltd, Bristol (1996)

Calcium metabolism in birds and mammals

F Bronner

Department of BioStructure and Function, University of Connecticut Health
Center, Farmington, CT 06030-3705, USA

It was recently proposed (3) that acute plasma Ca regulation is mediated by Ca binding
sites in bone, with high-affinity binding sites occurring predominantly in the more
mature phases of the bone mineral, whereas low-affinity sites are associated with the
phases of forming bone mineral (Table 1). From Ca loading experiments in rats and
other mammals, the rate of disappearance of the load following its initial dispersion in
the extracellular fluid (3, 5) has been found to have a $t_{1/2}$ of tens of minutes (3) and
therefore the plasma Ca returns to the preload level quite rapidly. This is true
regardless of the amount of the load (5) and applies also to the disappearance of
radiocalcium from the plasma and its uptake by bone (3). Moreover, when
radiocalcium passes over bone, about half is taken up, even when the total plasma Ca
level remains unchanged (3, 5). When EDTA is injected, the plasma Ca being lowered,
return to the preinjection plasma Ca level occurs at the same rate as when a positive
load is given (5, 8). Therefore each time blood courses over bone, half of the positive
difference between the plasma Ca level at that instant and the final plasma Ca is
abolished. Consequently the plasma Ca seems to be in rapid, dynamic equilibrium with
Ca binding sites in bone. Whereas such an equilibrium is generally considered in terms
of the solubility product of ionic Ca and phosphate in the plasma, an equilibrium can
equally be described in terms of an apparent half-saturation concentration, $K\mu$.
Moreover, when there is no net Ca gain or loss from the plasma, the free energy of Ca
binding must be zero and the plasma Ca level, $[Ca_s]$, must equal the mean $K\mu$ of the
binding sites.

Table 1 lists possible binding sites in bone with their varying $K\mu$ values. From
what has been said above, it is apparent that the mean $K\mu$ of all sites must equal and
determine the $[Ca_s]$. More correctly, it is the ionic Ca concentration in the plasma that
equals the $K\mu$, but because there normally is a close correlation between $[Ca_s]$ and
$[Ca^{2+}]$ (2), it is permissible to say $K\mu=[Ca_s]$.

In order to vary $K\mu$ and therefore $[Ca_s]$, the proportion of high-affinity and
low-affinity binding sites must be changed. This is accomplished, it has been proposed
(3), by shape changes of the metabolically active bone cells, i.e. the osteoblasts and
osteoclasts. If osteoblasts are primarily associated with bone surfaces where Ca
deposition is initiated and osteoclasts with surfaces where bone mineral has matured to
hydroxyapatite, then shape changes will free up low-affinity binding sites previously
covered up by extended osteoblasts or high-affinity sites covered up by extended

Table 1 Solid phases of calcium phosphate linked to biological calcification. (Listed in order of increasing acidity, solubility, and, by inference, of increasing Kμ, i.e. decreasing calcium binding affinity.) Reproduced by permission from (17). Note: at least 90% of the crystalline solid is hydroxyapatite.

Name	Formula	Molar Ratio, Ca/P
Hydroxyapatite	$Ca_{10}(PO_4)_6(OH)_2$	1.66
Whitlockite	$(CaMg)_3(PO_4)_2$	1.50
Amorphous Ca_xPO_4	$Ca_9(PO_4)_6$(variable)	1.30-1.50
Octacalcium phosphate	$Ca_8H_2(PO_4)_6 \cdot 5H_2O$	1.33
Brushite	$CaHPO_4 \cdot 2H_2O$	1.00

osteoclasts. Hormone interaction with cell receptors of bone cells have been shown to bring about rapid shape changes in osteoblasts (11, 15) and osteoclasts (1, 6). Osteoblasts have receptors for parathyroid hormone (PTH) and 1,25-dihydroxyvitamin D_3 (1,25-$(OH)_2$ D_3) (3, 15) and for calcitonin (CT) (6). Rodan and Martin (14) proposed, on the basis of experimental findings subsequently widely confirmed, that shape changes in osteoblasts lead to opposite shape changes in osteoclasts. Thus when the osteoblasts round up in response to PTH, osteoclasts have an opportunity to extend and when osteoclasts round up in response to CT, osteoblasts can extend. In response to PTH, therefore, the exposure of more low-affinity sites and the covering up of high-affinity sites leads to a rise in Kμ and [Ca_s]. The rounding up of osteoclasts in response to CT will expose more high-affinity sites, the corresponding extension of osteoblasts will cover up more low-affinity sites. As a result, CT administration will result in a lowering of Kμ and of [Ca_s]. This is illustrated in Fig. 1.

Approximately 5% of the resting cardiac output of a dog flows to the skeleton (12, 13) and since half of the ^{45}Ca that enters the circulation of a dog's femur is extracted from the blood (12), one can calculate that it would require 27 circulations (ln 0.5/ln 0.975) to reduce a circulating load to half. This would mean 18 min in the dog, circulation time 40 s, and 9 min in the rat, circulation time 20 s (18), numbers clearly consistent with what has been found (3). Moreover, the bone clearance rate is a function of cardiac output and circulation time, both of which are age-dependent (18), and the degree and extent of skeletal mineralization, which are affected by age, hormonal and nutritional status. The increase in clearance times in dogs (20) and in vitamin D-deficient rats (5) can be explained (3) on the basis of changes in these three factors, i.e. circulation time, cardiac output and extent of mineralization.

Most of the experimental studies on which the above hypothesis is based were done in mammals, although rapid changes in bone Ca uptake following PTH administration have also been documented in chick bone (16). What may seem surprising is the relative lack of response of birds to CT (7, 9). The process of shell

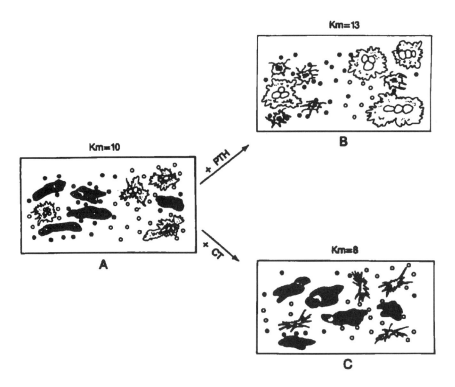

Fig. 1 Diagram representing the effect of the acute administration of PTH or CT on the spatial relationships of osteoblasts and osteoclasts on a bone surface. The open and closed circles represent respectively the Kμ values of the high-affinity and low-affinity bone calcium-binding sites. Diagram A represents a normocalcemic situation, with an equal number of high- and low-affinity sites. Diagram B represents the result of PTH administration, where the shrinkage of osteoblasts has exposed low-affinity sites and the associated expansion of osteoclasts has blocked high-affinity sites, leading to an average Kμ of 13 and hypercalcemia. Diagram C represents the result of CT administration, where the shrinkage of osteoclasts has exposed high-affinity sites and the consequent expansion of osteoblasts has blocked low-affinity sites, leading to an average of Kμ of 8 and hypocalcemia. Note: high-affinity sites are considered to have an apparent Kμ of 5 mg Ca/dl, with low-affinity sites having an apparent Kμ of 15 mg Ca/dl. For the sake of convenience, the Kμ values refer to total plasma Ca, rather than the theoretically correct ionic Ca concentration, approximately half of the total. Bone mineral with a high Ca/P ratio is considered to have a higher affinity for Ca binding than bone mineral with a low Ca/P ratio (see text and Table 1). Reproduced with permission from (3).

formation takes place in a matter of hours (10) and requires significant amounts of Ca. While birds can bind substantial amounts of Ca in the plasma with the aid of proteins (19), thereby minimizing a corresponding rise in ionized Ca, the amount of protein must nevertheless go up, as Ca input into the plasma increases. If this were to lead to a marked rise in CT and therefore an increase in the amount of Ca directed to bone, shell calcification might be interfered with.

Hurwitz and Bar (10) showed that in the course of calcification, the amount of absorbed Ca goes up fourfold, with much of this increase occurring in the jejunum which does not contain calbindin D-28k, the vitamin D-dependent protein involved in transcellular transport (4). If, as seems likely, most of the increase in jejunal Ca absorption is via the paracellular pathway (4), the transient increase in plasma Ca must be great. If bone cells of chickens were as exquisitely sensitive to CT as mammalian osteoclasts, the rise in plasma $[Ca^{2+}]$ would provoke rapid release of CT and increased Ca flow to bone instead of to the calcifying shell. This prediction would be strengthened if an association were to be found between receptor sensitivity to CT or the number of CT receptors in chicken bone cells, degree of egg shell calcification and, possibly, egg production.

References

1   Boyde A & Jones SJ (1987) Early scanning electron microscopic studies of hard tissue resorption: Their relation to current concepts reviewed. *Scanning Microscopy* **1** 369-381.

2   Bronner F (1982) Calcium homeostasis. *Disorders of Mineral Metabolism,* pp 43-97. Eds F Bronner & JW Coburn. New York: Academic Press.

3   Bronner F & Stein WD (1992) Modulation of bone calcium-binding sites regulates plasma calcium: an hypothesis. *Calcified Tissue International* **50** 483-489.

4   Bronner F, Pansu D & Stein WD (1986) An analysis of intestinal calcium transport across the rat intestine. *American Journal of Physiology* **250** (*Gastrointestinal Liver Physiology* **13**) G561-569.

5   Bronner F, Bosco JJ & Stein WD (1989) Acute plasma calcium regulation in rats: effect of vitamin D deficiency. *Bone and Mineral* **6** 141-154.

6   Chambers TJ, McSheehy PMH, Thompson BM & Fuller K (1985) The effect of calcium-regulating hormones and prostaglandins on bone resorption by osteoclasts disaggregated from neonatal rabbit bones. *Endocrinology* **60** 234-239.

7   Clark NB & Wideman RF Jr (1980) Calcitonin stimulation of urine flow and sodium excretion in the starling. *American Journal of Physiology* **238** (*Regulatory Integrative and Comparative Physiology* **7**) R406-R412.

8   Copp DH, Moghadam H, Mensen ED & McPherson GD (1961) The parathyroids and calcium homeostasis. In *The Parathyroids,* pp 203-219. Eds RO Greep & RV Talmage. Springfield, IL: CC Thomas.

9   Hurwitz S (1989) Calcium homeostasis in birds. *Vitamins and Hormones* **45** 173-221.

10   Hurwitz S & Bar A (1990) Oscillations in intestinal calcium absorption as observed and as predicted by computer stimulation. In *Calcium Transport and Intracellular Calcium Homeostasis,* pp 257-262. Eds D Pansu & F Bronner. Berlin: Springer-Verlag.

11   Miller SS, Wolf AM & Arnaud CD (1976) Bone cells in culture: morphologic transformation by hormones. *Science* **192** 1340-1342.

12   Ray RD (1976) Circulation and bone. In *The Biochemistry and Physiology of Bone*, 2nd edn, pp 385-402. Ed GH Bourne. New York: Academic Press.

13   Ray RD, Kawabata M & Galante J (1967) Experimental studies of peripheral circulation and bone growth. *Clinical Orthopaedics and Related Research* **54** 175-185.

14   Rodan GA & Martin TJ (1981) Role of osteoblasts in hormonal control of bone resorption –a hypothesis. *Calcified Tissue International* **33** 349-351.

15   Sato M & Rodan GA (1989) Bone cell shape and function. In *Cell Shape Determinants, Regulation and Regulatory Role*, ch 11, pp 329-362. Eds WD Stein & F Bronner. New York: Academic Press.

16   Shaw AJ & Dacke CG (1989) Cyclic nucleotides and the rapid inhibitions of bone ^{45}Ca uptake in response to bovine parathyroid hormone and 16,16-dimethyl prostaglandin E_2 in chicks. *Calcified Tissue International* **44** 209-213.

17   Simmons DJ & Grynpas MD (1990) Mechanisms of bone formation *in vivo*. In *Bone: vol 1, The Osteoblast and Osteocyte*, ch 6, table 4. Ed BK Hale. Caldwell, NJ: Telford Press.

18   Spector WS (Ed) (1956) *Handbook of Biological Data,* p 584, table 55. Philadelphia: WB Saunders Co.

19   Urist MR, Schjeide AO & McLean FC (1958) The partition and binding of calcium in the serum of the laying hen and of the estrogenized rooster. *Endocrinology* **63** 570-585.

20   Waron M & Rich C (1969) Rate of recovery from acute hypocalcemia as a measure of calcium homeostatic efficiency in the dog. *Endocrinology* **85** 1018-1027.

The Comparative Endocrinology of Calcium Regulation
Eds C Dacke, J Danks, I Caple & G Flik, pp 137-141
Journal of Endocrinology Ltd, Bristol (1996)

Response of the calcium-regulating system in chickens to growth

S Hurwitz and A Bar

Institute of Animal Science, Agricultural Research Organization, The
Volcani Center, Bet Dagan, Israel

Background

Plasma calcium (Ca) concentration is one of the most guarded physiological qualities of higher animals. The system of feedback-regulation of plasma Ca includes several components, with either rapid or slow response times (Fig. 1). The initial step in regulation is Ca sensing by a specific receptor (2).

Factors which affect plasma Ca but are not linked by any feedback mechanism to plasma Ca, are not considered as part of the Ca control system and are therefore not included in Fig. 1.

The Ca control system responds to perturbations in plasma Ca by modulating Ca flows. Rapid responses are either directly associated with the Ca concentration, such as the kidney flow, or with the action of peptide hormones, most importantly parathyroid hormone (PTH). During a chronic perturbation such as a prolonged reduction in Ca intake, the rapid PTH release and the resulting non-economic increase in bone resorption rate is replaced by the more sluggish but more economic increase in Ca absorption in response to 1,25-dihydroxycholecalciferol $(1,25(OH)_2D_3)$, the production of which is also stimulated by PTH.

Growth of body mass including that of bone may be considered as a steady state perturbation of the Ca regulatory system, since it involves a proportional outflow of Ca from the central pool. The rate of growth changes as a function of age, genetic potential and nutritional-environmental factors. Therefore, it may be expected that at least some of the components of the Ca-regulating systems would respond to these factors.

Due to the homeostatic regulation of plasma Ca, its concentration is poorly correlated with the nutritional status or the rates of movement of Ca between body compartments such as soft tissue and bone. These can be studied either by use of isotopes such as ^{45}Ca or by introducing specific perturbations. System perturbations such as parathyroidectomy or vitamin D deficiency have been used to evaluate the importance of the associated hormones in Ca homeostasis. However, these drastic approaches result in a large departure of metabolism from a 'normal' status and in chicks can actually lead to death. Other approaches include the use of bolus Ca loading or chronic feeding of diets of different Ca concentrations. A computer simulation algorithm may be of importance in integrating specific knowledge gained from whole animals or cellular experimentation and may enable the prediction of the complex sequences of events which follow any perturbation of the control system.

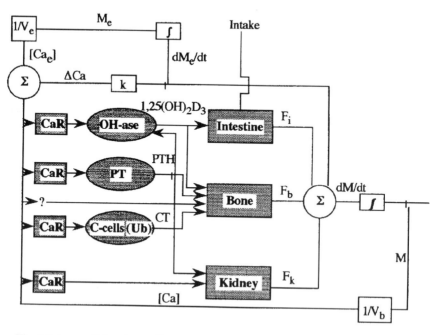

Fig. 1 Scheme of the plasma Ca regulation system. Plasma Ca concentration [Ca] is monitored by Ca-sensing receptors (CaR) identified on cells of parathyroid (PT), kidney and in C cells of the thyroid gland of most mammals and the ultimobranchial glands (Ub) of birds, and probably also on some bone cells. These control systems respond to CaR by modulating their activity. Parathyroid gland varies secretion of parathyroid hormone (PTH), the ultimobranchial gland secretes calcitonin (CT), kidney cells produce 1,25-dihydroxyvitamin D_3 (1,25(OH)$_2$D$_3$) by modulating the activity of 25-cholecalciferol-1-hydroxylase (OH-ase) and also vary urinary Ca excretion (F_k), and bone cells modulate Ca apposition and release. The produced hormones modify Ca flows in bone (F_b), intestine (F_i) and kidney (F_k). [Ca] is then the integral of the Ca flows divided by the plasma volume (V_b) which is in equilibrium with interstitial Ca, [Ca$_e$]. [Ca$_e$] is the result of total interstitial Ca (M_e) divided by the intestitial fluid volume (V_e). Most of the relationships described in the scheme have been modeled in a computer algorithm (7).

Effect of growth on the capacity to regulate plasma Ca concentration

The pattern of response of plasma Ca to dietary Ca concentration (Fig. 2) is reminiscent of buffer titration. In the chick, a steady plasma Ca concentration is maintained within the wide range of dietary Ca (0.8-1.5%). An intact vitamin D system is essential for this remarkable regulatory ability (6).

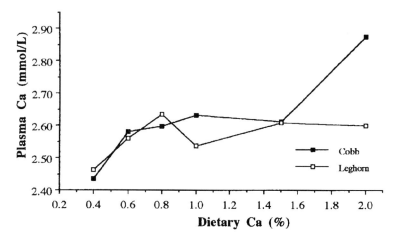

Fig. 2 The response of plasma Ca of fast- (Cobb) and slow- (Leghorn) growing chickens to dietary Ca concentration.

In the growing chick, the Ca regulation capacity is overwhelmed when challenged with intakes below this range, regardless of growth rate. However, an increase in plasma Ca at levels above 1.5% is observed only in the rapidly growing and not in the slow-growing chick, suggesting a difference between the two types of birds with regard to the Ca regulation capacity.

Effect of growth on components of the Ca-regulating system

The age-related or nutritionally induced growth retardation is associated with a decreased activity of the parathyroid gland, a low activity of the 25-hydroxyvitamin D_3-1-hydroxylase and a low concentration of intestinal calbindin which is linearly correlated with the efficiency of Ca absorption (1). The same variables are lower in slow-growing than rapidly growing breeds (1, 9). The lower activity may be explained on the basis of the feedback relationships shown in Fig. 1. However, the amount of maximal suppressible intestinal calbindin from dietary calcium remained higher in the rapidly-growing than in slow-growing birds, suggesting that factors outside of the Ca-control system may also be involved in modulation of system components by the process of growth (9).

Oscillations in the Ca regulating system

Oscillations in plasma Ca and several components of the Ca-regulating system manifested as diurnal variations, have been observed in rat (15) dog (16) humans (10, 11) and recently in the growing chicken (8). Although there is no agreement on the exact identity of the source of the these oscillations, both Hurwitz *et al.* (7) and Staub *et al.* (15) suggested that they are a property of the Ca-regulating system and are related to its normal function, rather than being a result of system errors. Furthermore, such oscillations may be of benefit to the organism (12). Computer simulation suggested that the oscillations were the result of the different response times of the two systems activated by PTH-bone resorption and vitamin D-intestinal Ca absorption. Furthermore, oscillations appeared proportional in amplitude and frequency to the rate of growth and in effect almost disappeared when simulated growth was abolished (7).

Ca nutriture and growth

The need for an adequate dietary supply of Ca to support the maximal growth rate of rats was documented as early as the 1930s by Sherman and Campbell (14). In chickens, Gardiner (3) observed a reduced growth rate in chickens when dietary Ca was reduced below 0.4%. The comparison of slow- and rapidly growing chickens (9) showed that the growth rate of the slow-growing birds either was not affected by dietary Ca or was slightly retarded at the low range of dietary Ca. In the rapidly growing birds, growth retardation occurred in addition at high Ca intakes, in agreement with Shafey et al. (13), so that a bell-shaped function of growth on dietary Ca was obtained (Fig. 3).

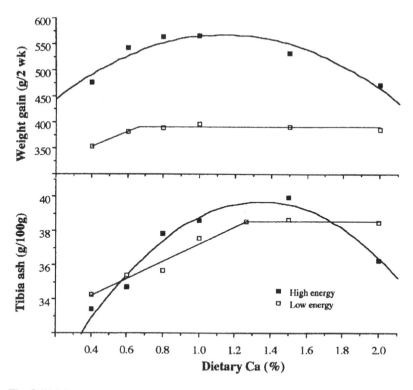

Fig. 3 Weight gain (upper graph) and bone calcification (lower graph) of rapidly growing (high-energy) and slow-growing (low-energy) chickens.

Similarly to growth rate, bone calcification in rapidly growing chickens (Fig. 3) changed with dietary Ca in a bell-shaped pattern, whereas the bone mineral of the slow-growing birds was reduced only at low dietary calcium intake. The reason for growth and bone calcification inhibition by high Ca intakes has not been elucidated but may be due to a hypophosphatemia (9), caused by suppression of intestinal phosphate absorption (4, 5). This, however, does not explain the absence of response of the slow-growing birds.

References

1   Bar A & Hurwitz S (1981) Relationship between cholecalciferol metabolism and growth in chicks as modified by age, breed and diet. *Journal of Nutriton* **111** 399-404.

2   Brown EM, Gamba G, Riccardi D, Lombard M, Butters R, Kifor O, Sun A, Hediger MA, Lytton J & Herbert FC (1993) Cloning and characterization of an extracellular Ca^{2+}- sensing receptor from bovine parathyroid. *Nature* **366** 575-580.

3   Gardiner EE (1971) Calcium requirements of breeds of chickens as influenced by levels of dietary phosphorus. *Canadian Journal of Animal Science* **51** 445-450.

4   Hurwitz S & Bar A (1971) Calcium and phosphorus interrelationships in the intestine of the fowl. *Journal of Nutrition* **101** 677-685.

5   Hurwitz S, Dubrov D, Eisner U & Riesenfeld G (1978) Phosphate absorption and excretion in the young turkey, as influenced by calcium intake. *Journal of Nutrition* **108** 1329-1335.

6   Hurwitz S, Fishman S, Bar A & Talpaz H (1984) Regulation of calcium absorption by 1,25 dihydroxycholecalciferol and its importance in calcium homeostasis. *Transfer of Calcium and Strontium Across Biological Membranes*, pp 357-362. Eds F Bronner & M Peterlik. New York: Allen Liss Inc.

7   Hurwitz S, Fishman S & Talpaz H (1987) Model of plasma calcium regulation: systems oscillations induced by growth. *American Journal of Physiology* **252** R1173-R1181.

8   Hurwitz S, Miller B, & Norman AW (1994) Oscillatory behavior of control-systems of calcium homeostasis in chickens. *Journal of Cellular Biochemistry* **56** 236-244.

9   Hurwitz S, Plavnik I, Shapiro A, Wax E, Talpaz H & Bar A (1995) Calcium metabolism and requirements of chickens are affected by growth. *Journal of Nutrition* (In press).

10   Jubitz W, Canterbury JM, Reiss E & Tyler F (1972) Circadian rhythms in serum parathyroid hormone concentration in human subjects: correlation with serum calcium, phosphate, albumen and growth hormone levels. *Journal of Clinical Investigation* **52** 2040-2046.

11   Markowitz M, Rotkin L & Rosen JF (1981) Circadian rhythms of blood minerals in humans. *Science* **213** 672-674.

12   Moore-Ede MC (1986) Physiology of the circadian timing system: predictive *versus* reactive homeostasis. *American Journal of Physiology* **250** R735-R752.

13   Shafey TM, McDonald MW,& Pim RAE (1990) The effect of dietary calcium upon growth rate, food utilisation and plasma constituents in lines of chickens selected for aspects of growth or body composition. *British Poultry Science* **31** 577-586.

14   Sherman HC & Campbell HL (1936) Effect of increasing the calcium content of a diet in which calcium is one of the limiting factors. *Journal of Nutrition* **8** 363-371.

15   Staub JF, Tracqui P, Brezillon P, Milhuad G & Perault-Staub A-M (1988) Calcium metabolism in the rat: a temporal self-organized model. *American Journal of Physiology* **254** R134-R149.

16   Wong KM & Klein L (1984) Circadian variations in contributions of bone and intestine to plasma calcium in dogs. *American Journal of Physiology* **246** R688-692.

The Comparative Endocrinology of Calcium Regulation
Eds C Dacke, J Danks, I Caple & G Flik, pp 143-148
Journal of Endocrinology Ltd, Bristol (1996)

Ability of quail medullary osteoclasts *in vitro* to form resorption pits and stain for tartrate resistant acid phosphatase

Z A Bascal and C G Dacke

School of Pharmacy and Biomedical Sciences, University of Portsmouth,
PO1 2DT, UK

In egg-laying birds, such as the domestic hen and Japanese quails, there is an almost daily demand for high levels of calcium (Ca) for egg calcification. In order to meet these high requirements, female birds, upon reaching sexual maturity, develop medullary bone in the marrow cavities of their long bones which has been regarded as a Ca reservoir supplying as much as 40% of the Ca necessary for egg shell formation (for a review see (7)). The high turnover of medullary bone composition during a 24 h egg laying cycle (10), which is under hormonal control, suggests that medullary bone osteoclasts are highly regulated and are responsive to ionic changes in the extracellular fluid. In addition, the destruction of medullary bone by the withdrawal of sex hormones is analogous to what occurs in human post-menopausal osteoporosis. Thus, medullary bone osteoclasts could provide an attractive model for studying and characterizing osteoclastic behaviour in response to changes in hormonal and humoral factors as occurs in osteoporosis.

The mode of action of potential calciotropic agents on bone resorption can be assessed either in whole animal systems, organ cultures or in isolated cell cultures. Isolated cell cultures require the seeding of bone cells, preferably pure osteoclastic cultures, onto pieces of devitalized bone. Bone resorption is then monitored by evaluating resorption pit volume and area with scanning electron microscopy (6, 9) or light microscopy (12).

For the present study, we investigated the ability of cultured osteoclasts obtained from medullary bone of egg-laying Japanese quails, to resorb bone, using the pit formation assay, and to stain for tartrate-resistant acid phosphatase (TRAP), as two markers for characterizing osteoclasts *in vitro* (6, 11).

Methods

Osteoclasts were isolated from medullary bone of Japanese quail hens (*Coturnix coturnix japonica*) as described by Bascal (3). Briefly, hens aged 4-10 months which had been kept on a 16 h light:8 h darkness cycle and fed on either a quail egg-laying diet (rich in Ca and vitamin D) or on Ca and a vitamin D deficient diet (for 3-4 weeks), were fasted overnight (water available *ad libitium*) and selected on the basis that they contained a palpable egg. After cervical dislocation, their femurs and tibias were removed and the medullary contents were scraped from the bone cylinders into

(a)   ├───┤ 20 μm

(b)   ├───┤ 20 μm

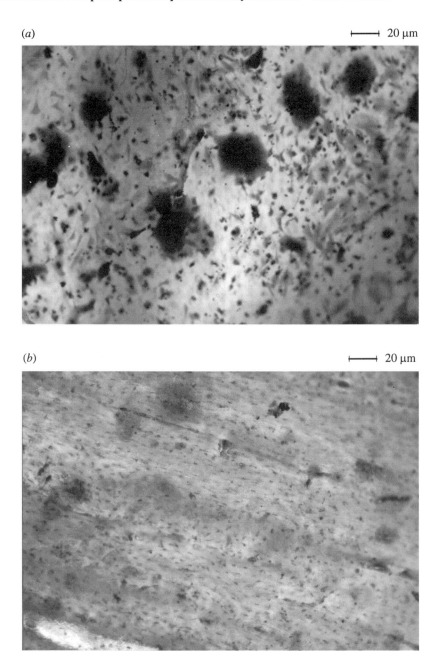

Fig. 1 (a) Light micrograph of osteoclast-like cells settled onto bone slices for seven days and stained for the presence of TRAP (dark stain). Scale bar 20 μm. (b) Bone slices preincubated for seven days stripped of cells and stained for TRAP. Slices show stained patches which may correspond to resorption pits.

Hepes-buffered medium 199, chopped and agitated followed by filtration (350 μm mesh). Freshly prepared bone cell suspension was aliquoted onto pre-wetted devitalized bovine cortical bone slices (0.4 cm^2 in area and 0.1 mm in thickness) arranged in 24 well plates, containing Medium 199 and 10% fetal calf serum (FCS). Osteoclasts were allowed to sediment and attach to the substrate for 2-4 h. Following incubation, cell cultures were washed in minimal essential medium (MEM) containing 10% FCS, penicillin (100 iu/ml), streptomycin (100 μg/ml) at 37.5 °C, 10% humidified CO_2 for 24 h. The cultures were subsequently washed daily and kept for up to seven days. For comparative studies, osteoclasts were isolated from neonatal rats and newly hatched quail long bones, using similar isolation procedures as reported elsewhere (2, 4). Bone slices were then either (1) fixed in acetone-citrate buffer and stained for TRAP or (2) stripped of their cells by washing in distilled water containing Triton X 100 (0.1%) and processed for scanning electron microscopy, to examine osteoclastic excavations. For detail of procedures see Bascal (3).

Results

Seven day old medullary bone cultures contained numerous cells of varying sizes and shapes. The devitalized bone slices were heavily populated with cells, many of which stained positively for TRAP (Fig. 1), indicating their osteoclast-like characteristics. Negatively stained cells were often mononucleated and of smaller size than the osteoclast-like cells that stained for TRAP. However, numerous small mononuclear cells, which are likely to be osteoclast precursors, also stained for TRAP. Various patches on bone slices also stained positively for TRAP. These patches may represent osteoclastic pits, where secreted TRAP has accumulated, as they tend to be in close proximity to osteoclast-like cells (Fig. 1(a)). Such results are similar to those reported elsewhere for chick osteoclasts (8). These patches are more prominent following the removal of cultured cells, prior to staining (Fig. 1(b)).

Examination of bone slices incubated with cells from either neonatal rat or newly hatched quail, long bones, or from female quail medullary bone, revealed the presence of pits, indicating bone resorption (Figs 2(a)-(d)). The pits were of varying size and shape irrespective of the source of osteoclasts. However, the pits produced by medullary bone osteoclast cultures tended to be fewer in number per bone slice and tended to vary greatly in size, shape and depth. In particular numerous large but very shallow pits were produced that could only be observed upon tilting the specimens in the electron microscope. Such shallow depressions were not seen with either neonatal rat osteoclast cultures or with newly hatched quail cultures. The incubation of bone slices with medullary bone osteoclasts obtained from quails maintained on a Ca and vitamin D deficient diet for three weeks, had no overall effect on the number or types of pits produced. In addition, alterations in the culture conditions, such as incubation time, concentration of FCS and CO_2, did not appear to influence pit formation by medullary bone osteoclasts.

(*a*)

(*b*)

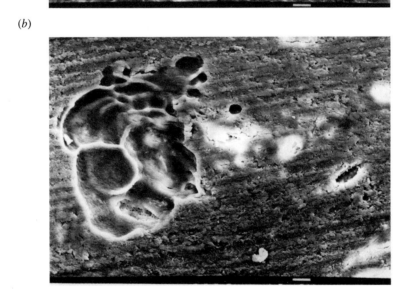

Fig. 2 Scanning electron micrographs of bone slices incubated with: (*a*) neonatal rat osteoclasts. Removal of cells reveals the presence of classical resorption pits with well defined boundaries. Scale bar 10 μm. (*b*) Osteoclasts from newly hatched quails. The resorption pits revealed are essentially similar to those formed by rat osteoclasts. (*c*) Medullary bone osteoclasts. Note the large well defined but irregular shape of these pits Scale bar 10 μm. (*d*) Medullary bone osteoclasts. Numerous shallow pits of this type, which reveal depth upon tilting, are observed. Note the minor excavations surrounding the main patch. Scale bar 100 μm.

(c)

(d)

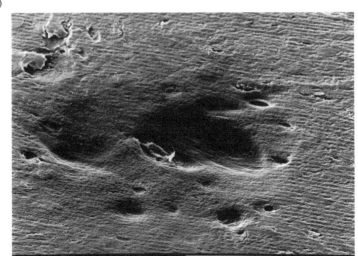

Discussion

Using a simple isolation technique, cells that possess osteoclast-like characteristics can be isolated and cultured from Japanese quail medullary bone. The cells are large, multinucleated, stain positively for TRAP and resorb bone by pit formation. Additionally the cells secrete TRAP (3, 7) and respond to increases in extracellular Ca^{2+} by increasing their intracellular Ca^{2+} and contracting (1, 5). These features confirm the osteoclastic nature of the cells isolated from Japanese quail meduliary bone.

Osteoclasts isolated from mammalian and newly hatched chicks and quails attach very rapidly and with strong anchorage to bone substrate, allowing vigorous washing of slices to remove non-osteoclastic cells. However, medullary bone osteoclasts, require 6-12 h to adhere, thus enabling other non-osteoclastic cells to attach to the substrate. This results in a mixed culture of cells and overcrowding which may explain the low number of pits obtained with medullary bone osteoclasts in comparison to neonatal osteoclast cultures. The very shallow but large depressions observed appear unique to medullary bone osteoclasts. This feature could be a reflection of the highly complex chemical and physical organization of medullary bone with its honeycomb-like structure which provides a preferential substrate for local osteoclasts.

References

1   Arkle S, Wormstone IM, Bascal ZA & Dacke DG (1994) Estimation of intracellular calcium activity in confluent monolayers of primary cultures of quail medullary bone osteoclasts. *Experimental Physiology* **79** 975-982.

2   Arnett TR & Dempster DW (1987) A comparative study of disaggregated chick and rat osteoclasts *in vitro:* effects of calcitonin and prostaglandins. *Endocrinology* **120** 602-608.

3   Bascal ZA (1994) PhD Thesis. University of Portsmouth. UK.

4   Bascal ZA, Moonga BS, Zaidi M & Dacke DG (1992) Osteoclasts from medullary bone of egg-laying Japanese quail do not possess the putative calcium receptor. *Experimental Physiology* **77** 501-504.

5   Bascal ZA, Alam ASMT, Zaidi M & Dacke DG (1994) Effect of raised extracellular calcium on cell spread area in quail medullary bone osteoclasts. *Experimental Physiology* **79** 15-24.

6   Chambers TJ, Revell PA, Fuller K & Athanasou NA (1984) Resorption of bone by isolated rabbit osteoclasts. *Journal of Cell Science* **66** 383-399.

7   Dacke CG, Arkle S, Cook DJ, Wormstone IM, Jones S, Zaidi M & Bascal ZA (1993) Medullary bone and avian calcium regulation. *Journal of Experimental Biology* **184** 63-88.

8   Hunter SJ, Rosen CJ & Gay CV (1991) *In vitro* resorptive activity of isolated chick osteoclasts: effects of carbonic anhydrase inhibition. *Journal of Bone and Mineral Research* **6** 61-66.

9   Jones SJ, Boyde A & Ali NN (1984) The resorption of biological and non-biological substrates by cultured avian and mammalian osteoclasts. *Anatomy and Embryology* **170** 247-256.

10   Miller SC (1977) Osteoclast cell-surface changes during the egg-laying cycle in Japanese quail. *Journal of Cell Biology* **75** 104-118.

11   Minkin C (1982) Bone acid phosphatase: tartrate-resistant acid phosphatase as a marker of osteoclast function. *Calcified Tissue International* **34** 285-290.

12   Sato M & Grasser W (1990) Effects of biphosphonates on isolated rat osteoclasts as examined by reflected light microscopy. *Journal of Bone and Mineral Research* **5** 31-40.

The Comparative Endocrinology of Calcium Regulation
Eds C Dacke, J Danks, I Caple & G Flik, pp 149-150
Journal of Endocrinology Ltd, Bristol (1996)

Morphological changes of osteoclasts on hen medullary bone during the egg-laying cycle and their regulation

T Sugiyama and S Kusuhara

Department of Animal Science, Faculty of Agriculture, Niigata University,
2-8050 Ikarashi, Niigata 950-21, Japan

Medullary bone is specific to female birds and serves as a reservoir of labile calcium to supplement dietary calcium for egg-shell formation. Osteoclasts on the medullary bone change in their ultrastructure and bone resorptive ability during the egg-laying cycle (1, 2). However, their regulating hormones have not been clarified. In the present study, we first observed the morphological behaviour of medullary bone osteoclasts during the egg-laying cycle and then tried to clarify the regulating hormones using organ culture systems of medullary bones.

At 0 to 6 h after oviposition, osteoclasts lacking ruffled borders attached to the bone surface via clear zones and appeared to have ceased bone resorption. Small vacuoles were scattered throughout the cytoplasm. On the other hand, at 9 to 21 h after oviposition, most of the osteoclasts had ruffled borders and appeared to be resorbing bone. The ruffled borders at 15 h were well developed, whereas at 9, 18 and 21 h they were poorly developed or showed similarities in structure to the clear zone. The small vacuoles were concentrated under the ruffled borders at 9 h and then decreased. Secondly, medullary bones at 3 h after oviposition were cultured in the presence of human parathyroid hormone (PTH) (5 U/ml medium) (Asahikasei Co. Ltd, Tokyo, Japan), and medullary bones at 12 h after oviposition were cultured in the presence of eel calcitonin (CT) (50 mU/ml medium) (Asahikasei Co. Ltd, Tokyo, Japan). Their frozen sections were stained with nitrobenzoxadiazole (NBD)-phallacidin, actin filaments were observed using a fluorescent microscope and the ultrastructure of osteoclasts was observed. At 3 h after oviposition, osteoclasts did not possess ruffled borders and attached to the medullary bone via clear zones, and actin filaments were localized as amorphous bands at the apical region of the cytoplasm, just where clear zones exist. On the other hand, at 12 h after oviposition, osteoclasts with developed ruffled borders were observed and actin filaments were localized as lines perpendicular to the bone surface, corresponding to the positions of ruffled borders. In the presence of PTH, ruffled borders appeared and became well developed, and linear actin filaments were observed. In the presence of CT, ruffled borders disappeared and osteoclasts had actin filaments as amorphous bands at the apical region of the cytoplasm.

These morphological changes suggest that osteoclastic bone resorption begins at 9 h after oviposition, peaks at 12 to 15 h and decreases thereafter. This indicates that ruffled borders derive from the small vacuoles and the clear zone, and that the ruffled borders fuse into the clear zone and disappear at the completion of bone resorption. Moreover, it is concluded that PTH stimulates osteoclastic resorption of the medullary bone, while CT inhibits. Therefore, PTH and CT regulate osteoclastic resorption of the medullary bone during the egg-laying cycle.

References

1   Miller SC (1977) Osteoclast cell-surface changes during the egg- laying cycle in Japanese quail. *Journal of Cell Biology* **75** 104-118.
2   Miller SC (1981) Osteoclast cell-surface specializations and nuclear kinetics during egg-laying in Japanese quail. *American Journal of Anatomy* **162** 35-43.

The Comparative Endocrinology of Calcium Regulation
Eds C Dacke, J Danks, I Caple & G Flik, pp 151-154
Journal of Endocrinology Ltd, Bristol (1996)

Unusual (pathological?) histology in an *Iguanodon* caudal centrum

J B Clarke and M J Barker

Department of Geology, University of Portsmouth, Burnaby Road,
Portsmouth PO1 3QL UK

Introduction

Histological details displayed by an *Iguanodon* (Dinosauria: Ornithischia) caudal vertebral centrum, MIWG 7320, were observed to be unusual. Three other centra were examined one of which, IWGMS:1994:14, has been used for comparison.

Overall trabeculae patterns in transverse and longitudinal sections of the normal specimen, display a regular pattern (Fig. 1). The central cancellous zone is composed of a regular arrangement of trabeculae surrounded by roughly equi-sized voids. Some voids in the ventral half of the centrum are elongated radially and one or two radially orientated nutrient canals are also present. As the cortex is approached the voids become smaller. Transverse sections from the pathological centrum are shown in Fig. 2. Although some post-mortem crushing has occurred, much of the internal structure is still intact and clearly visible. In the dorsal half of the centrum the voids show a similar pattern to the normal centrum, although without the same degree of regularity, but in the ventral half, all the voids are radially deposed and are markedly elongated. Six radial nutrient canals are also visible in the specimen.

Histology

In the authors' experience, the pathological bone shows unusual histological detail. Cortical bone at the sides of normal centra is usually parallel-fibred, sometimes with resting lines, with a reasonable density of concentrically arranged vascular canals, while the cortex at the ventral end is sometimes primary woven bone, containing primary osteons. The vascularity of the cortex in the pathological specimen is unusually high and, together with the presence of woven bone, suggests a high growth rate. The radial cortical patterns in the ventral end of the centrum is also highly unusual and again the presence of woven bone together with the direction of linearity further suggest abnormally high growth rates.

Inside the cortex of iguanodontid vertebral centra, secondary Haversian rings are normally rare, and there is usually a transition zone where primary bone is resorbed and secondary bone deposited. This gradual change is characterised by small resorption holes which progressively become bigger towards the centre of the centrum. In the pathological specimen, large resorption spaces directly invade the cortex; there is no transition zone where small resorption holes occur.

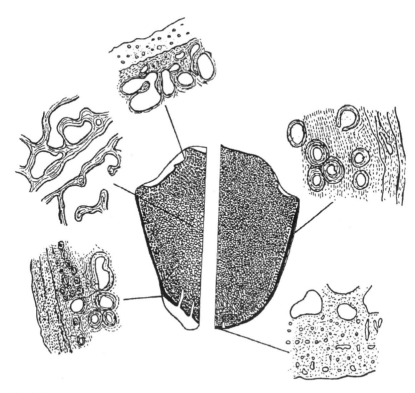

Fig. 1 Transverse section of the normal vertebral centrum showing sketches of the overall trabeculae patterns and histological details (not to scale).

In the normal iguanodontid centra, the trabeculae in the cancellous zone usually contain a high percentage of secondary, endosteal bone, remodelling having continued until only a thin core (if any) of primary bone remains. Sometimes primary vascular canals are retained in the outer areas of the cancellous zone, but in the centre of the centrum most have been removed by remodelling. In the pathological specimen, the trabeculae throughout most of the cancellous zone are composed of unaltered primary vascular bone surrounded by a thin coating of endosteal bone; there are only small areas in which the trabeculae are composed of secondary endosteal bone. The thick patches of primary bone in the centre of of this centrum are highly unusual; trabeculae in this region are usually thin.

Discussion

It is unfortunate that only a single skeletal element was available for study. However, the comparative histologies described above clearly indicate that the bone growth and modification in the pathological centrum was unusual. Whilst clearly very speculative, the possible cause of these differences raises a number of intriguing scenarios.

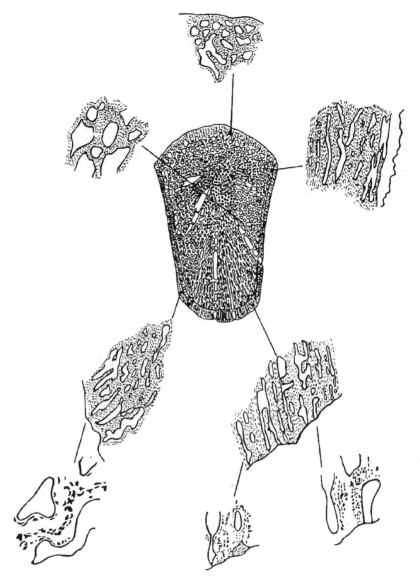

Fig. 2 Transverse section of the pathological vertebral centrum showing sketches of the overall trabeculae patterns and histological details (not to scale).

The alternative aetiologies are: (a) local trauma inducing single-element pathology; (b) disease affecting the whole organism; (c) genetically inherited abnormalities; (d) haemangioma (a form of neoplasm). Most pathological studies are based upon mammal physiology and therefore any comparisons can only be tentative.

All features in the pathological centrum suggest that bone deposition and resorption were abnormally rapid whilst remodelling and deposition of secondary bone was very slow or absent.

(a) **Local trauma** One possibility could be localised trauma during early ontogeny causing low-grade osteoperiostitis or traumatic arthropathy.

(b) **Disease** The bone histology is reminiscent of osteoporosis in mammals, described as 'decreased bone formation, increased bone resorption or a combination of both' (1). However, the high rate of growth of primary bone is more akin to Paget's disease (*Osteitis deformans*), which is also known to attack only single bone elements. Histological illustrations of Paget's disease show a remarkable resemblance to the dorsal end of the pathological specimen. A further speculative cause could be vitamin-related (deficiency or otherwise) if the condition was pervasive throughout the organism.

(c) **Genetic causes** There is, as yet, no means of relating genetic abnormalities to bone histology in dinosaurs. We have no observed evidence to suggest any genetic cause for this condition, although some form of endocrine abnormality remains a possibility.

(d) **Haemangioma** It has been suggested that the trabecular pattern found in haemangiomas is similar to that in the *Iguanodon* and this diagnosis is being investigated by use of X-rays (2).

Conclusions

Clearly further systematic surveys to ascertain details of dinosaur bone histology and variation are required. Much of the thin sectioning of fossil bones to date has been concerned with the ectothermy/endothermy debate, but also random sections have been cut to compare with modern mammals and reptiles and to describe the detailed histology of dinosaurian bones. Exactly where each type of histology occurs within each bone, the variation between individual bones and the variation between species has yet to be ascertained. However, the pathological specimen described above, shows enough variation from the normal to be considered unusual if not malignant; the growth structures indicating that something was stimulating bone growth to an abnormal degree.

Acknowledgements

The authors wish to thank S Hutt of the Museum of the Isle of Wight for providing the specimen and Dr D Bradwell for discussion on the pathology.

References

1   Cappell DF (1964) *Muir's Textbook of Pathology.* 8th edn. London: Edward Arnold.
2   Rothschild BM & Martin LD (1992) *Paleopathology: Diasease in the Fossil Record.* Boca Raton, FL: CRC Press.

Part Three

Mammals

The Comparative Endocrinology of Calcium Regulation
Eds C Dacke, J Danks, I Caple & G Flik, pp 157-164
Journal of Endocrinology Ltd, Bristol (1996)

Endocrine control of mammalian calcium metabolism

I W Caple

University of Melbourne, Department of Veterinary Science, Veterinary
Clinical Centre, Princes Highway, Werribee, Victoria 3030, Australia

Introduction

The extracellular ionised calcium concentrations in adult mammals are regulated
within a narrow range of 1.15-1.35 mmol/l to maintain normal muscular contraction,
neural excitability, blood coagulation, enzyme activity, hormone release and
membrane permeability. Calcium homeostasis involves interactions between three
major hormones: parathyroid hormone (PTH), calcitonin, and 1,25 dihydroxyvitamin
D (1,25(OH)$_2$D), the hormone of the vitamin D endocrine system (9). Plasma Ca
concentrations below this range are detected by a calcium-sensing receptor on the
parathyroid cell surface that is coupled to intracellular second messenger systems
through one or more guanine nucleotide regulatory (G) proteins (8). The parathyroid
cells are rapidly triggered to secrete PTH which restores normocalcaemia by
increasing Ca mobilisation from bone, increasing tubular reabsorption of Ca in the
distal nephron and stimulating proximal tubular synthesis of 1,25(OH)$_2$D, which
increases absorption of Ca from the intestine (16). Elevated extracellular Ca
concentrations above this range, or hypercalcaemia, stimulate the thyroid C cells to
secrete calcitonin and inhibit PTH secretion by the parathyroid cells. The increased
1,25(OH)$_2$D concentrations may also affect parathyroid function directly by inhibiting
expression of the PTH gene. The major calciotropic hormones also have roles in other
physiological functions in adult mammals, but these will not be considered in this brief
review. An additional calciotropic hormone, parathyroid hormone-related protein
(PTHrP), appears to be important during pregnancy and foetal development (22, 25,
26) and its physiological actions such as promoting placental calcium transport,
maintaining foetal hypercalcaemia, and contributing to foetal skeletal mineralisation
and remodelling will be reviewed.

Challenges to calcium homeostasis in adult mammals

Pregnancy, lactation and skeletal growth present major challenges to calcium nutrition
and homeostasis in adult mammals. Intestinal Ca absorption is markedly increased in
humans during pregnancy and lactation and bone mineral stores are preserved (17).
Some ruminant mammals, particularly sheep, differ from monogastic mammals in that
absorption of calcium in excess of requirements rarely occurs (7). Clinical
hypocalcaemia (plasma Ca<1.7 mmol/l) accompanied by general muscular weakness,

recumbency and depression of consciousness is encountered more frequently in domesticated cows, ewes and does than in any other mammalian species.

The reliance on the dairy cow for milk has led to a sixfold increase in milk production by elite cows during the twentieth century through genetic selection, nutritional management and improved disease control. Whereas the 2400 l of milk containing 2.8 kg Ca produced by a 400 kg cow in 1906 was equivalent to half the cow's total body calcium (5.6 kg Ca), the 15 000 litres of milk containing 18 kg Ca produced by a 600 kg cow in 1995 represents twice her total body Ca. The increased demands on the cow's homeostatic mechanisms has also resulted in an increased prevalence of clinical hypocalcaemia occurring around the time of parturition and onset of lactation.

In pregnant ewes, hypocalcaemia occurs when there is insufficient absorption of Ca from the small intestine and resorption from skeletal reserves to meet the demands of the foetus (up to 2.8 g Ca/day) (18). The growth of the foetus and its consequent demands for Ca play the major role in defining the rate of placental transfer of Ca and the Ca requirements of the pregnant ewe. Irrespective of their Ca intake, ewes do not absorb sufficient dietary Ca to meet the requirements for placental Ca transfer to the developing foetal skeleton during pregnancy and for milk secretion in early lactation (7). As a consequence, ewes may mobilise up to 20% of skeletal Ca reserves which are replaced in mid to late lactation when Ca demands for milk production fall. Calcitonin may protect the skeleton of pregnant and/or lactating ewes from excessive demineralisation (5, 6). Plasma osteocalcin, a product of osteoblasts and a measure of bone formation, was found to decrease from day 35 and remain low during pregnancy in ewes (15). Plasma osteocalcin then increased from day ten of lactation to levels above those in nonpregnant control ewes, reflecting increased bone formation to restore the depleted skeletal reserves (15).

Calcium homeostasis in the mammalian foetus

In all mammals that have been studied, the foetus is hypercalcaemic relative to its mother's plasma Ca concentration during the latter stages of gestation (4, 13, 17). The timing of the establishment of this placental Ca gradient during gestation has not been examined in many mammalian species. Our studies in sheep have shown that foetal plasma ionised and total Ca concentrations are maintained higher than maternal levels from very early in gestation (10, 25). In this species, active placental Ca transfer is established by day 35, soon after the foetal superior parathyroid glands have appeared and the attachment of the trophoblast to the maternal caruncles is complete. Calcification of the skeleton commences after the establishment of the gradient, and is first observed in the mandible in the foetal lamb. The relative hypercalcaemic state of the foetus is established before the period of placental growth from days 35 to 90, and is maintained during the period of rapid foetal growth and bone development from day 90 to birth (145-150 days). In the sheep foetus, almost all Ca that passes to it from the mother is used by the developing skeleton (18). Placental Ca transport in sheep is essentially a unidirectional process, and there is no net resorption of Ca from foetal bone (7).

Role of the foetal parathyroid glands in foetal Ca metabolism

The parathyroid glands, which develop from the third and fourth brachial pouches, appear relatively earlier in gestation in those mammalian species whose young are born in a more advanced stage of skeletal development and mineralisation. In sheep they appear during the fourth week (20% of gestation), during the eighth week in man (22% of gestation) and second week in rats (60% of gestation) (26). Little is known about the development of the parathyroid glands in the marsupials such as kangaroos, koalas and wallabies which are born in near embryonic state and complete their development attached to their mothers' teats.

The sheep is a useful model to study foetal calcium metabolism as experiments can be conducted on ewes with twin pregnancies to known conception dates (12). Thyroparathyroidectomy (TXPTX) of one twin foetus was carried out between days 90 and 120 of gestation. Catheters filled with heparinised saline were inserted into a foetal carotid artery and jugular vein under halothane anaesthesia. Following surgery, ewes were kept in single pens and fed lucerne and oaten chaff, lucerne hay and oats *ad libitum*. Each day the catheters were filled with heparinised saline and 20 ug thyroxine was injected into the jugular vein to maintain the TXPTX foetus in a euthyroid state. Blood samples were collected from the jugular vein of the ewe and the carotid artery of the foetus for estimation of plasma ionised and total Ca.

This experimental preparation has been used to monitor the effects of foetal TXPTX on plasma Ca (10, 12, 25), placental Ca transport (12), bone formation (1, 10), renal function (24) and the responses to intravenous infusions of Ca, bovine parathyroid hormone (bPTH)(1-34), bPTH(1-84), human (h)PTHrP(1-34), recombinant hPTHrP(1-141), and of several PTHrP mid-molecule fragments (2, 14, 25). Foetal parathyroid glands were extracted and examined for PTH-like activity in bioassays (3, 25). Some were fixed in formol saline before being embedded in paraffin, sectioned and examined histochemically for the presence of ovine PTHrP using antisera raised against fragments of hPTHrP (22).

The maintenance of the hypercalcaemia in the foetal lamb is clearly dependent on the foetal parathyroid glands because TXPTX leads to a rapid reversal of the Ca gradient across the placenta within two to three days (12, 26). The resultant hypocalcaemia is attributed to a decreased activity of the placental Ca pump (12) and a reduced rate of Ca transfer to the foetus. The bones of hypocalcaemic TXPTX foetuses show decreased calcification of cartilage and mineralisation of osteoid (1).

Detection of PTHrP in the foetus

Until the discovery of PTHrP, the foetal parathyroid gland factors responsible for maintaining foetal Ca homeostasis were unknown. Circulating concentrations of immunoreactive PTH were undetectable in foetal lamb plasma and infusions of PTH at physiological concentrations to TXPTX lambs could not restore hypercalcaemia (11). Several observations led to the hypothesis that a parathyroid-hormone-like protein may be the factor responsible for maintaining the activity of the placental Ca pump. Plasma PTH-like bioactivity measured by a sensitive cytochemical assay was higher in the foetus than in the mother and this bioactivity could not be inhibited by the presence of an N-terminally directed antiserum to bPTH (3, 4). Biological assays of extracts of foetal lamb parathyroid glands and sheep placenta provided further evidence for the existence of a PTHrP which could not be neutralised by an antiserum against PTH (21, 25).

Immunohistological localisation of PTHrP in the chief cells of the foetal lamb parathyroid gland has been made using paraffin sections and a peroxidase-antiperoxidase method with antisera directed to a fragment of the mid-molecule (50-69) region of hPTHrP (22).

Effect of PTHrP on placental Ca transport

A physiological role for PTHrP in the foetus was obtained in a series of experiments carried out on the placentae of TXPTX and intact twin foetal lambs (2, 25). The placentae were perfused with autologous foetal blood after the foetus was removed at the end of pregnancy and we measured the ability of the placenta to pump Ca against a concentration gradient (12). In TXPTX lambs the activity of the placental Ca pump was markedly reduced. Addition of bPTH(1-84), and synthetic N-terminal(1-34) peptides of bPTH and hPTHrP to the foetal blood perfusing the placenta had no effect on the placental Ca pump (25). However, extracts of foetal parathyroid glands, and purified and recombinant hPTHrP(1-141) and hPTHrP fragments (1-84 and 1-108) stimulated placental Ca transport (2, 14, 25). This led to the hypothesis that placental activity of the hPTHrP molecule does not reside in the N-terminal region, which shows homology with PTH. As hPTHrP(1-84), hPTHrP(1-108), and PTHrP(67-86 amide) showed a more rapid stimulatory action on the placenta than extracts of foetal para-thyroid glands or hPTHrP(1-141), it has been suggested that some post-translational processing of PTHrP may be required as an activating step in the placenta (2, 14).

Foetal bone formation

The changes in foetal bone formation during gestation, after foetal TXPTX and during infusions of Ca, bPTH, hPTHrP were monitored by measuring changes in plasma alkaline phosphatase activity and osteocalcin concentration. Plasma hydroxyproline was measured as an indicator of bone resorption. In addition, chemical and histomorphometric measurements were made on sections of lumbar vertebrae of TXPTX and intact twin foetuses.

The changes in plasma osteocalcin and alkaline phosphatase throughout gestation reflected the increases in skeletal growth (11). Plasma alkaline phosphatase activity rose from a mean of 90 IU/l at 60 days of gestation to 350 IU/l at term. Plasma osteocalcin increased from 52.5 µg/l ($n=2$) at 60 days reaching 371 ± 64.2 µg/l (mean±SEM, $n=23$) in the last month. While studies with Ca radioisotopes have indicated there is no net resorption of bone Ca in the foetus (7), we observed that plasma hydroxyproline concentrations doubled between day 67 and day 114 of gestation then remained relatively constant at a mean of 71.0 ± 2.6 µmol/l in the last month. We interpreted this observation to indicate that there was remodelling of bone without net Ca resoprtion during foetal growth.

Effect of foetal TXPTX on foetal calcaemia and bone formation

TXPTX of foetal lambs ($n=29$) resulted in rapid decreases in plasma Ca (from 3.22±0.12 to 1.95±0.10 mmol/l). Within two to three days after TXPTX, foetal plasma osteocalcin concentrations decreased from 229±66 to 43±6 µg/l after seven days, due to a reduced production of osteocalcin without changes in osteocalcin clearance (11). Plasma alkaline phosphatase decreased from 180±38 to 125±14 IU/l, and hydroxyproline decreased from 55.6±3.8 to 36.8±3.6 µmol/l. Histomorphometric analysis of sections of the lumbar vertebrae of intact and twin TXPTX foetuses indicated that foetal TXPTX and prolonged hypocalcaemia was associated with decreased calcification of cartilage and mineralisation of oeteoid (1). There were also reductions in the numbers of osteoblasts (from 80 to 45 mm^{-2} surface area) and osteoclasts (from 20 to 2 mm^{-2} surface area). However, it was not known whether these changes were due to the removal of a foetal parathyroid factor such as PTH which is known to stimulate osteoblastic activity and osteoclastic resorption, or to the prolonged hypocalcaemia following foetal TXPTX (11).

Effects of restoration of foetal hypercalcaemia on bone formation

The importance of hypercalcaemia for foetal bone formation was studied by comparing the changes in plasma osteocalcin, alkaline phosphatase, and hydroxyproline in TXPTX lambs, with those in TXPTX lambs given infusions of Ca into the jugular vein to restore foetal hypercalcaemia. Histomorphometric and chemical measurements were made on transverse sections of the fourth lumbar vertebrae of intact and twin TXPTX lambs 14 to 30 days after TXPTX (1), and in TXPTX twin foetuses where foetal hypercalcaemia was maintained by infusion for up to 21 days.

Infusions of Ca to TxPTX foetuses at 460 mg Ca/day (approximately 40% of placental Ca transport) were required to restore plasma Ca concentrations similar to those before foetal TXPTX. Foetal osteoblast activity appeared to be very sensitive to changes in foetal plasma Ca concentration. Plasma osteocalcin and alkaline phosphatase increased within 6 h after Ca was infused to TXPTX foetuses. The maximum concentrations reached during the 'Ca clamp' were similar to those before foetal surgery (osteocalcin: 281 and 229 µg/l respectively; alkaline phosphatase: 313 and 212 IU/l respectively). Plasma hydroxyproline concentration which had decreased from 44.5 to 12.7 µmol/l following foetal TXPTX, increased to 36 µmol/l. While chemical and histomorphometric analyses indicated mineralisation of bone in TXPTX foetuses was increased by Ca infusion, bone resorption was less than in intact twin foetuses, and the numbers of osteoblasts (53 mm^{-2}) and osteoclasts (<1 mm^{-2}) were still less than those in intact foetuses (99 and 20 mm^{-2} respectively). From these observations it seemed that osteoblastic activity and foetal bone formation were dependent on a high plasma Ca concentration, but that other factors were required for recruitment of osteoblasts and osteoclasts and bone remodelling.

Effect of PTH and PTHrP on foetal bone

Our initial studies showed that the placental Ca gradient could not be restored in TXPTX foetuses when they were given a bolus of bPTH(1-84) of up to 1.6 nmol/kg followed by a continuous infusion of up to 0.6 nmol/kg/h for 7 h. This was not unexpected since PTH is normally undetectable in foetal plasma, and PTH had no effect on Ca transport across the perfused placenta of TXPTX foetuses (25). Following the discovery of hPTHrP and reports that hPTHrP(1-34) stimulated bone resorption (26), we investigated whether higher doses of bPTH and hPTHrP were able to promote net bone resorption and restore hypercalcaemia in TXPTX foetal lambs in which the bones were labelled with ^{45}Ca. Within 10 h of infusing either bPTH(1-34), hPTHrP(1-34) or (1-141) there were detectable increases in the foetal plasma Ca concentration and specific activity of ^{45}Ca, and foetal hypercalcaemia was returned to preoperative levels within 24 h. The increases in plasma specific ^{45}Ca activity following infusion of hPTHrP(1-34) and (1-141), and bPTH(1-34), were similar, indicating that the N-terminal ends of bPTH and hPTHrP act on bone either directly or indirectly via the kidney through stimulation of increased plasma 1,25(OH)$_2$D, or by both mechanisms. Plasma 1,25(OH)$_2$D concentrations in TXPTX foetuses increased during infusion of either bPTH(1-34) or hPTHrP(1-34) and (1-141) (11) and also increased in intact foetal lambs following injection of these peptides (5).

Calcium homeostasis and bone formation in the lamb

The blood ionised and total calcium concentrations in lambs are normally maintained higher than in the ewe for up to six months after birth and during the period when there is most skeletal growth. PTHrP may play a role in calcium homeostasis and bone formation in lambs after birth. It can be demonstrated to be present in the parathyroid glands of lambs up to six months of age by immunohistochemical techniques, but not in older lambs and mature sheep (22). We postulate that PTHrP may maintain plasma Ca at the expense of bone quality in lambs whenever milk and calcium intake is inadequate. Evidence for this was obtained in an experiment in which Merino lambs were fed different intakes of milk from one week of age and allowed access to pasture (19). Lambs which grew less than 100 g/day in the first six weeks had osteoporosis and femoral fractures when autopsied at 12 weeks of age. The calcium requirements of lambs for appetite, growth, and maintenance of Ca homeostasis are much lower than for maximum mineralisation of the developing skeleton (250 versus 450 mg Ca/kg liveweight/day) (20). The calcium concentrations in herbage diets are unlikely to limit appetite and growth of lambs, but may limit skeletal mineralisation. In flocks where ewes have insufficient milk for their lambs in the early neonatal period, osteoporosis with fractures of the femur and ribs may occur in the lambs.

Conclusion

Studies during the past decade have shown that foetal hypercalcaemia is maintained by a factor produced in the foetal parathyroid gland. Evidence is accumulating that the factor is a PTHrP. PTHrP shares many functions with PTH in its action on the kidney, bone and the vitamin D endocrine system. The N-terminal (1-34) regions of both hormones have a similar sequence of amino acids. In the foetus, PTHrP also stimulates the placental transfer of Ca, and this function appears to be related to a mid-molecule region of PTHrP and is not shared with PTH. PTHrP produced by the foetal parathyroid gland and elsewhere, such as the placenta and foetal membranes, may have a multifunctional role in foetal Ca metabolism to ensure that elevated foetal plasma Ca concentrations are maintained and bones are adequately mineralised during rapid foetal growth. The high foetal plasma Ca concentration is maintained relatively constant throughout gestation by the placental Ca pump under the control of PTHrP from the foetal parathyroid gland. Bone formation in the foetal lamb is dependent on a high foetal plasma concentration, but it appears that the presence of humoral factors provided by the thyroparathyroid complex is also required for bone remodelling. In foetal plasma, PTH is undetectable but the high levels of PTHrP may contribute to bone remodelling during rapid skeletal growth. It is likely that the placental activity of PTHrP is determined by the C-terminal portion of the molecule. The N-terminal region of PTHrP, which is similar to PTH, appears to be responsible for its actions on bone, the kidney and the vitamin D system.

References

1   Aaron JE, Makins NB, Caple IW, Abbas SK, Pickard DW & Care AD (1989) The parathyroid glands in the skeletal development of the ovine foetus. *Bone and Mineral* **7** 13-22.

2   Abbas SK, Pickard DW, Rodda CP, Heath JA, Hammonds RG, Wood WI, Caple IW, Martin TJ & Care AD (1989) Stimulation of ovine placental calcium transport by purified natural and recombinant parathyroid hormone-related protein (PTHrP) preparations. *Quarterly Journal of Experimental Physiology* **74** 549-552.

3   Abbas SK, Pickard DW, Illingworth D, Storer J, Purdie DW, Moniz C, Dixit M, Caple IW, Ebeling PR, Rodda CP, Martin TJ & Care AD (1990) Measurement of parathyroid hormone related protein in extracts of fetal parathyroid glands and placental membranes. *Journal of Endocrinology* **124** 319-325.

4   Abbas SK, Ratcliffe WA, Moniz C, Dixit M, Caple IW, Silver M, Fowdem A & Care AD (1994) The role of parathyroid hormone-related protein in calcium homeostasis in the fetal pig. *Experimental Physiology* **79** 527-536.

5   Barlet J-P (1985) Calcitonin may modulate placental transfer of calcium in ewes. *Journal of Endocrinology* **104** 17-21.

6   Barlet J-P, Davicco M-J & Coxam V (1992) Calcitonin modulates parathyroid hormone related peptide stimulated calcium placental transfer. In *Calcium Regulating Hormones and Bone Metabolism,* pp 124-128. Eds DV Cohn, C Gennari & AH Tashjian Jr. Amsterdam: Excerpta Medica.

7   Braithwaite GD & Glascock RF (1976) Metabolism of calcium in sheep. *Biennial Reviews of the National Institute for Research in Dairying* **4425** 43-59.

8   Brown EM, Pollak M & Hebert SC (1995) Molecular mechanisms underlying the sensing of extracellular Ca^{2+} by parathyroid and kidney cells. *European Journal of Endocrinology* **132** 523-531.

9   Capen CC & Rosol TJ (1989) Calcium-regulating hormones and diseases of abnormal mineral (calcium, phosphorus, magnesium) metabolism. *Clinical Biochemistry of Domestic Animals,* pp 678-752. Ed JJ Kaneko. San Diego: Academic Press.

10   Caple IW, Heath JA, Care AD, Heaton C, Farrugia W & Wark JD (1988) The regulation of osteoblast activity in fetal lambs by placental calcium transport and the fetal parathyroid glands. *Fetal and Neonatal Development,* pp 90-93. Ed CT Jones. Ithaca, NY: Perinatology Press.

11   Caple IW, Heath JA, Pham TT, Farrugia W, Wark JD, Care AD & Martin TJ (1990) The role of the parathyroid glands, PTH, PTHrP and elevated plasma calcium in bone formation in fetal lambs. *Calcium Regulation and Bone Metabolism: Basic and Clinical Aspects,* vol 10, pp 455-460. Eds DV Cohn, FH Glorieux & TJ Martin. Amsterdam: Excerpta Medica.

12   Care AD, Caple IW, Abbas SK & Pickard DW (1986) The effect of fetal thyroparathyroidectomy on the transport of calcium across the ovine placenta to the fetus. *Placenta* **7** 417-424.

13   Care AD, Caple IW, Singh R & Peddie M (1986) Studies of calcium homeostasis in the fetal Yucatan miniature pig. *Laboratory Animal Science* **36** 389-392.

14   Care AD, Abbas SK, Pickard DW, Barri M, Drinkhill M, Findlay JBC, White IR & Caple IW (1990) Stimulation of ovine placental transport of calcium and magnesium by mid-molecule fragments of human parathyroid hormone-related protein. *Experimental Physiology* **75** 605-608.

15   Farrugia W, Fortune CL, Heath J, Caple IW & Wark JD (1989) Osteocalcin as an index of osteoblast function during and after ovine pregnancy. *Endocrinology* **125** 1705-1710.

16   Fraser DR (1995) Vitamin D. *The Lancet* **345** 104-107.

17   Garel J-M (1987) Hormonal control of calcium metabolism during the reproductive cycle in animals. *Physiological Reviews* **67** 1-66.

18   Grace ND, Watkinson JH & Martinson PL (1986) Accumulation of minerals by the foetus(es) and conceptus of single- and twin-bearing ewes. *New Zealand Journal of Agricultural Research* **29** 207-222.

19   Heath JA & Caple IW (1988) Importance of milk intake in preventing osteoporosis in young lambs at pasture. *Proceedings of the Nutrition Society of Australia* **13** 88.

20   Hodge RW, Pearce GR & Tribe DE (1973) Calcium requirements of the young lamb: 1 effects of different intakes of dietary calcium on liveweight gain, bone development and blood serum calcium levels. *Australian Journal of Agricultural Research* **24** 229-236.

21   Loveridge N, Caple IW, Rodda C, Martin TJ & Care AD (1988) Evidence for a parathyroid hormone-related protein in fetal parathyroid glands of sheep. *Quarterly Journal of Experimental Physiology* **73** 781-784.

22   MacIsaac RJ, Caple IW, Danks JA, Diefenbach-Jagger H, Grill V, Moseley JM, Southby J & Martin TJ (1991) Ontogeny of parathyroid hormone-related protein in the ovine parathyroid gland. *Endocrinology* **129** 757-764.

23   MacIsaac RJ, Heath JA, Rodda CP, Moseley JM, Care AD, Martin TJ, & Caple IW (1991) Role of the fetal parathyroid glands and parathyroid hormone-related protein in the regulation of placental transport of calcium, magnesium and inorganic phosphate. *Reproduction, Fertility and Development* **3** 447-457.

24   MacIsaac RJ, Caple IW, Martin TJ & Wintour EM (1993) Effects of thyroparathyroidectomy, parathyroid hormone, and PTHrP on kidneys of ovine fetuses. *American Journal of Physiology* **264** (*Endocrinology and Metabolism* **27**) E37-E44.

25   Rodda CP, Kubota M, Heath JA, Ebeling PR, Moseley JM, Care AD, Caple IW & Martin TJ (1988) Evidence for a novel parathyroid hormone-related protein in fetal lamb parathyroid glands and sheep placenta: comparisons with a similar protein implicated in humoral hypercalcaemia of malignancy. *Journal of Endocrinology* **117** 261-271.

26   Rodda CP, Caple IW & Martin TJ (1992) Role of PTHrP in fetal and neonatal physiology. In *Parathyroid Hormone-Related Protein: Normal Physiology and its Role in Cancer,* pp 169-196. Eds BP Halloran & RA Nissenson. Boca Raton, FL: CRC Press Inc.

The Comparative Endocrinology of Calcium Regulation
Eds C Dacke, J Danks, I Caple & G Flik, pp 165-169
Journal of Endocrinology Ltd, Bristol (1996)

Effect of parathyroid hormone-related protein on renal calcium excretion in the intact and thyroparathyroidectomized ovine fetus

R J MacIsaac[1], I W Caple[2], T J Martin[3] and E M Wintour[1]

[1]Howard Florey Institute of Experimental Physiology and Medicine, University of Melbourne, Parkville, Victoria 3052, Australia; [2]Department of Veterinary Science, University of Melbourne, Werribee, Victoria 3030, Australia and [3]St Vincent's Institute of Medical Research, 41 Victoria Parade, Fitzroy, Victoria 3065, Australia

Introduction

The ovine fetus is maintained hypercalcemic relative to its mother by a placental calcium (Ca) pump (3). Evidence is accumulating which suggests that parathyroid hormone-related protein (PTHrP) and not parathyroid hormone (PTH) may be the main factor responsible for maintaining elevated plasma Ca concentrations in the fetus (11). The fetal parathyroid glands contain both PTH-like bioactivity (14) and PTHrP immunoreactivity (2, 10); furthermore thyroparathyroidectomy (TXPTX) of the ovine fetus results in a reversal of the placental Ca gradient (5). Infusions of ovine fetal parathyroid gland extracts and various PTHrP but not PTH peptides were also found to increase the rate of Ca transport across the placenta of TXPTX ovine fetuses (2, 6, 14).

These findings led to our current hypothesis that a placental Ca pump, driven by PTHrP from the fetal parathyroid glands, maintains hypercalcemia in the fetus. Until recently, the role that fetal kidneys play in maintaining fetal plasma Ca concentrations had not been examined. In this paper we review our studies of the effects of TXPTX and PTHrP on the regulation of urinary Ca excretion in the ovine fetus.

Methods used to study renal Ca excretion in the fetus

Intact or TXPTX (with thyroxine replacement) ovine fetuses which had been chronically cannulated with indwelling silastic carotid, jugular and bladder cannulae were used to study fetal urinary Ca excretion. The fetal bladder was cannulated through an abdominal incision so as not to obstruct the normal passage of urine through the fetal urethra or urachus. Surgery was performed between 100 and 120 days of gestation (term is approximately 150 days) and animals allowed to recover for at least five days before any experiment was performed. Measurements of renal function were made in TXPTX or intact fetuses during a 1 h control period, a 2 h infusion period in which either saline or human PTHrP(1-34), PTHrP(1-141) or PTH(1-34) were infused, and a 1 h post-infusion period. In all experiments, peptides were infused at the rate of 1 nmol/h, after a 1 nmol loading dose. Measurements of parameters of urinary Ca excretion were also made in intact and Ca-infused TXPTX fetuses. TXPTX fetuses were infused with Ca, at rates of up to 20 mg/h, to restore the normally elevated plasma Ca concentrations, and hence the filtered loads of Ca, to values recorded for intact fetuses. These Ca-infused TXPTX fetuses then received infusions of saline or PTHrP(1-34) as described above. The filtered loads and fractional excretion rates of Ca were calculated assuming a ratio of ultrafilterable to plasma Ca of 0.67 (7). Details of the exact methodology and number of individual fetuses used in these studies have been documented previously (12, 13).

Comparison of renal Ca excretion in TXPTX and intact fetuses

In TXPTX ovine fetuses, fetal plasma Ca concentrations and the filtered loads of Ca (i.e. glomerular filtration rates × ultrafilterable Ca concentrations) were lower than values recorded for intact fetuses. However, even at these lowered filtered loads of Ca, the absolute excretion rates of Ca observed for TXPTX fetuses were not significantly different from those of intact fetuses. In relative terms, TXPTX fetuses must therefore excrete more Ca in comparison to intact fetuses. These differences were reflected by changes in the fractional excretion rates of Ca (calculated by dividing the absolute urinary excretion rates by the filtered loads of Ca) which were significantly elevated in TXPTX compared with intact fetuses (Table 1). Absolute excretion rates of 3′, 5′-cyclic monophosphate (cAMP) were also observed to be significantly lower in TXPTX fetuses than in intact fetuses. The results of these experiments would suggest that some factor secreted from the fetal parathyroid glands, presumably PTHrP, normally acts to inhibit the excretion of Ca by the fetal kidneys (12).

Table 1 Comparison of renal Ca excretion in intact and TXPTX fetuses. Abreviations: P_{Ca}, plasma Ca concentrations; FL_{Ca}, filtered loads of Ca; $U_{Ca}V$, absolute excretion rates of Ca; FE_{Ca}, fractional excretion rates of Ca. Results[a] are means±SEM with the number of fetuses studied shown in parenthesis. *$P<0.05$ (Mann-Whitney).

	P_{Ca} (mmol/l)	FL_{Ca} (mol/h)	$U_{Ca}V$ (mol/h)	FE_{Ca} (%)
Intact	2.8±0.1(6)	177±14(5)	8.0±1.7(6)	4.5±0.3(5)
TXPTX	1.8±0.1(5)*	111±20.3(5)*	10.3±2.5(5)	9.3±1.2(5)*

[a] Results adapted from MacIsaac *et al.* (12).

Effect of PTHrP infusions in TXPTX fetuses

The absolute urinary excretion rates of Ca significantly decreased, whilst those for cAMP significantly increased after infusions of PTHrP(1-34) into TXPTX fetuses. The fractional excretion rates of Ca and the ratio of the fractional excretion rate of Ca to the fractional excretion rate of sodium (a measure of Ca transport in the distal tubule) were also significantly decreased after infusions of PTHrP(1-34). Equimolar infusions of PTHrP(1-141) and PTH(1-34) produced similar changes in the above parameters of fetal renal function in comparison to the effects of PTHrP(1-34). These decreases in urinary Ca excretion were accompanied by significant increases in plasma Ca concentrations after infusions of all three PTHrP and PTH peptides (12). These results raise the question as to what are the mechanisms via which the fetus can maintain its plasma Ca concentrations. The 1-34 fragments of PTH and PTHrP are both capable of stimulating fetal bone resorption, but have no effect on placental Ca transport (1, 4). In contrast, PTHrP(1-141) could potentially stimulate Ca resorption from bone through its amino-terminal region, and its unique non-PTH-like mid-molecule sequence may be effective in stimulating Ca transport across the placenta from mother to fetus (6). Our results also suggests that the N-terminal region of PTHrP may contribute to the maintanence of fetal plasma Ca concentrations through the inhibition of urinary Ca excretion.

Effect of PTHrP infusions in intact fetuses

In contrast to TXPTX fetuses, infusions of PTHrP(1-34) into intact fetuses for 2 h had no significant effect on the urinary excretion rates of Ca and cAMP in comparison to control-saline infusions (9). We hypothesized that this difference in the response of intact and TXPTX fetuses to PTHrP was due to the fact that the filtered load of Ca presented to the renal tubules of intact fetuses exceeded the capacity of this dose of PTHrP to stimulate the reabsorption of Ca. To test this hypothesis we infused a group of TXPTX fetuses with Ca at a rate sufficient to elevate their plasma Ca concentrations and hence their filtered loads of Ca to levels equal to those recorded in intact fetuses. We then subsequently challenged these Ca-infused TXPTX fetuses with infusions of PTHrP(1-34).

Effect of PTHrP infusions in Ca-infused TXPTX fetuses

Although plasma Ca concentrations and the filtered loads of Ca in intact and Ca-infused TXPTX fetuses were not significantly different, the urinary concentrations, absolute excretion rates and fractional excretion rates of Ca were elevated in TXPTX fetuses (Table 2). To restore plasma Ca concentrations in TXPTX fetuses to values recorded for intact fetuses, Ca had to be infused at rates of up to 460 mg Ca/day (i.e. 10.5 mmol/day) as demonstrated previously by Caple et al. (3). The absolute excretion rate of Ca in these Ca-infused TXPTX fetuses was 32.0±13.3 mol/h (i.e. 0.77 mmol/day) which suggests that approximately 7.3% of the infused Ca was excreted in fetal urine. Subsequent infusions of PTHrP(1-34) into Ca-infused TXPTX fetuses, at an infusion rate that effectively inhibited urinary Ca excretion in TXPTX fetuses but not intact fetuses, resulted in significantly decreased urinary concentrations and absolute excretion rates of Ca over time. In comparison, control-saline infusions had no effect on Ca excretion over time (13). The results of these preliminary experiments confirmed the important role that the fetal parathyroid glands normally play in inhibiting urinary Ca excretion and suggested that the lack of effect of this infusion rate of PTHrP on Ca excretion in intact fetuses was not a function of the filtered load of Ca presented to the renal tubules. Davicco et al. (8) have also demonstrated that higher infusion rates of PTHrP(1-34), i.e. a 3 nmol bolus followed by a 3 nmol infusion for 30 min, into intact fetuses results in a significant hyperphosphaturia but has no effect on urinary Ca excretion. At present, the exact mechanism responsible for the difference in the renal Ca response of intact and TXPTX fetuses to PTHrP(1-34) infusions has yet to be elucidated.

Table 2 Comparison of renal Ca excretion in intact and Ca-infused TXPTX fetuses. Abreviations: P_{Ca}, plasma Ca concentrations; FL_{Ca}, filtered loads of Ca; $U_{Ca}V$, absolute excretion rates of Ca; FE_{Ca}, fractional excretion rates of Ca. Results[a] are means±SEM with the number of fetuses studied shown in parenthesis. *$P<0.05$, #$P=0.05$ (Mann-Whitney).

	P_{Ca} (mmol/l)	FL_{Ca} (mol/h)	$U_{Ca}V$ (mol/h)	FE_{Ca} (%)
Intact	2.8±0.15(5)	186±13(5)	6.9±2.2(5)	3.3±0.8(5)
Ca-infused				
TXPTX	2.8±0.2(5)	195±30(5)	32.0±13.3(5)#	14.6 4.5(5)*

[a] Results adapted from MacIsaac *et al.* (13).

Conclusion

The amino-terminal fragment of PTHrP is responsible for the actions of this peptide on Ca retention by the fetal kidney. This is in contrast to placental Ca transport, where mid-molecule fragments of PTHrP appear to be required to stimulate the transport of Ca from mother to fetus (6). In addition to regulating the activity of the placental Ca pump, the fetal parathyroid glands and PTHrP may also play an important role in inhibiting Ca excretion by the fetal kidneys to maintain hypercalcemia in the mammalian fetus.

Acknowledgements

This work was supported by grants from the NH & MRC of Australia to the Howard Florey Institute, The Anti-Cancer Council of Victoria and The Wool Research Trust Fund of The Australian Wool Corporation.

References

1   Abbas SK, Pickard DW, Rodda CP, Heath JA, Hammonds RG, Wood WI, Caple IW, Martin TJ & Care AD (1989) Stimulation of ovine placental calcium transport by purified natural and recombinant parathyroid hormone-related protein (PTHrP) preparations. *Quarterly Journal of Experimental Physiology* **74** 549-552.

2   Abbas SK, Pickard DW, Illingworth D, Storer J, Purdie DW, Moniz C, Dixit M, Caple IW, Ebeling PR, Rodda CP, Martin TJ & Care AD (1990) Measurement of parathyroid hormone-related protein in extracts of fetal parathyroid glands and placental membranes. *Journal of Endocrinology* **124** 319-325.

3   Caple IW, Heath JA, Care AD, Heaton C, Farrugia W & Wark JD (1988) The regulation of osteoblast activity in fetal lambs by placental calcium transport and the fetal parathyroid glands. In *Fetal and Neonatal Development,* pp 90-93. Ed CT Jones. Ithaca, NY: Perinatal Press.

4   Caple IW, Heath JA, Pham TT, MacIsaac RJ, Rodda CP, Farrugia W, Wark JD, Care AD & Martin TJ (1990) The roles of the parathyroid glands, parathyroid hormone (PTH), parathyroid hormone-related protein (PTHrP), and elevated plasma calcium in bone formation in fetal lambs. In *Calcium Regulation and Bone Formation,* pp 455-460. Eds DV Cohn, FH Glorieux & TJ Martin. Amsterdam/New York/Oxford: Excerpta Medica.

5   Care AD, Caple IW, Abbas SK & Pickard DW (1986) The effect of fetal thyroparathyroidectomy on the transport of calcium across the ovine placenta to the fetus. *Placenta* **7** 417-424.

6   Care AD, Abbas SK, Pickard DW, Barri M, Drinkhill M, Findlay JBC, White IR & Caple IW (1990) Stimulation of ovine placental transport of calcium and magnesium by mid-molecule fragments of human parathyroid hormone-related protein. *Experimental Physiology* **75** 605-608.

7   Carney SL, Wong NLM, Quamme GA & Dirks JH (1980) Effect of magnesium deficiency on renal magnesium and calcium transport in the rat. *Journal of Clinical Investigation* **65** 180-188.

8   Davicco MJ, Coxam V, LeFaire J & Barlet JP (1992) Parathyroid hormone-related peptide increases urinary phosphate excretion in fetal lambs. *Experimental Physiology* **77** 377-383.

9   Horne RSC, MacIsaac RJ, Moritz KM, Tangalakis K & Wintour EM (1993) Effect of arginine vasopressin and parathyroid hormone-related protein on renal function in the ovine fetus. *Clinical and Experimental Pharmacology and Physiology* **20** 569-577.

10   MacIsaac RJ, Caple IW, Danks JA, Diefenbach-Jagger H, Grill V, Moseley JM, Southby J & Martin TJ (1991) Ontogeny of parathyroid hormone-related protein in the ovine parathyroid gland. *Endocrinology* **129** 757-764.

11   MacIsaac RJ, Heath JA, Rodda CP, Moseley JM, Care AD, Martin TJ & Caple IW (1991) Role of the fetal parathyroid glands and parathyroid hormone-related protein in the regulation of placental transport of calcium, magnesium and inorganic phosphate. *Reproduction, Fertility and Development* **3** 447-457.

12   MacIsaac RJ, Horne RS, Caple IW, Martin TJ & Wintour EM (1993) Effects of thyroparathyroidectomy, parathyroid hormone and PTHrP on the kidneys of ovine fetuses. *American Journal of Physiology* **264** E37-E44.

13   MacIsaac RJ, Caple IW, Martin TJ, Moritz KM, Tangalakis K & Wintour EM (1994) Effect of calcium (Ca) infusion and parathyroid hormone-related protein (PTHrP) on renal Ca excretion in the thyroparathyoidectomized (TxPTx) ovine fetuses. *Society for the Study of Fetal Physiology Twenty First Annual Meeting* Abstract P20.

14   Rodda CP, Kubota M, Heath JA, Ebeling PR, Moseley JM, Care AD, Caple IW & Martin TJ (1988) Evidence for a novel parathyroid hormone-related protein in fetal lamb parathyroid glands and sheep placenta: Comparisons with a similar protein implicated in humoral hypercalcaemia of malignancy. *Journal of Endocrinology* **117** 261-271.

The Comparative Endocrinology of Calcium Regulation
Eds C Dacke, J Danks, I Caple & G Flik, pp 171-175
Journal of Endocrinology Ltd, Bristol (1996)

In vivo studies of the inhibitory effects of parathyroid hormone(1-34) and parathyroid hormone-related protein(1-34) on ruminal motility

J Harmeyer and A D Care[1]

Institute of Veterinary Physiology, Veterinary School, 3000 Hannover 1, Germany and [1]Institute of Biological Sciences, University of Wales Aberystwyth, Aberystwyth SY23 3DD, UK

Introduction

Parathyroid hormone (PTH) and parathyroid hormone-related hormone (PTHrP) share sufficient structural homology at their respective amino-termini to enable PTHrP to interact with classical PTH receptors in bone and kidney (5). PTHrP is the product of a gene that is expressed in many normal tissues. Its expression in the stomach can be induced by mechanical stretch (1). In addition, the expression of PTHrP mRNA in gastric smooth muscle of rats was shown to be markedly increased by gastric distension caused by pyloric ligation and also by the contractions induced by carbachol (2). It was further shown that PTHrP(1-34) relaxed cholinergically contracted gastric muscle strips, an effect shared with PTH(1-34), presumably through the common receptor they share (3).

Our work extends these studies to the cycle of contractions of the reticulo-rumen in conscious sheep using both exogenous PTH(1-34) or PTHrP(1-34) and endogenous PTH induced by a hypocalcaemic challenge.

Methods

Animals

A permanent fistula of diameter 12 cm was made in the dorsal rumen of 12 adult sheep (70-80 kg) and fitted with a soft plastic plug with inner and outer plastic flanges to prevent leakage of rumen contents. After at least two months, these sheep were used for the measurement of the changes in pressure in the reticulum, pre-rumen and dorsal and ventral posterior rumen, employing suitably weighted balloons connected via pressure transducers to a multi-channel recorder.

Experimental

(a) PTHrP(1-34)

After a control period of 10-44 min, during which the peptide vehicle (0.1 M NaCl, 0.01% bovine serum albumin) was infused i.v. into four sheep at the rate of 3.5 ml/min, a bolus i.v. injection of 12 µg PTHrP(1-34) was given. This was immediately followed by its i.v. infusion at 2.1 µg/min for 20 min. A further 40 min infusion of vehicle followed the end of the PTHrP(1-34) infusion. Samples of blood were taken for the assay of PTHrP(1-34). Measurements were made of the frequency and maximum amplitude of the biphasic reticular contractions, rumination and the frequency and amplitude of both the A and the B waves of contraction in the posterior dorsal sac and the B waves of contraction in the posterior ventral sac (as a verification).

(b) PTH(1-34)

A similar protocol was adopted with two sheep as used for PTHrP(1-34). Blood samples were taken to measure the concentrations of PTH(1-34) in the plasma.

(c) Control

The above measurements of contraction were made in two sheep over the same experimental period, during which no infusions were made.

(d) Hypocalcaemia

After an initial control period of i.v. saline infusion for 30-60 min, the endogenous secretion of PTH was stimulated in four sheep by the i.v. injection of 10 ml 0.15 M sodium citrate followed by its i.v. infusion at the rate of 2-4 ml/min for 1-2 h to reduce the plasma calcium ion (Ca^{2+}) concentration from the normal level of approximately 1.1 mmol/l to approximately 0.9 mmol/l. Before the sheep were returned to their pens, normocalcaemia was restored by the i.v. administration of 100 mM $CaCl_2$. This ensured that a normal pattern of reticulo-ruminal contractions was restored. Blood samples were taken at regular intervals to monitor the plasma Ca^{2+} concentration. Aliquots of plasma were rapidly frozen and stored at $-20\,°C$ until assayed for PTH(1-34).

To test the involvement of the PTH/PTHrP(1-34) receptor in reticulo-ruminal motility associated with i.v. citrate infusion, six experiments were carried out in four sheep which were first treated with the PTH/PTHrP(1-34) receptor blocker (Asn^{10}, Leu^{11} D-Trp^{12}) PTHrP(7-34) amide (4). This was injected i.v. as a bolus of 125 µg some 5 min before the start of the citrate and was followed by five i.v. injections of 12.5 µg at 10 min intervals, started 10 min after the first injection of the blocker. The same parameters were measured as above.

Analyses

Ca^{2+} concentrations were measured in heparinised blood under anaerobic conditions, using a specific Ca^{2+} electrode, shortly after the sample was taken. The plasma concentrations of PTH(1-34) were measured using a two-site assay kit (Nichols Institute, Saffron Walden, UK) based on rat PTH(1-34), which has a 30% cross reactivity with ovine PTH. The plasma levels of PTHrP(1-34) were measured with a radioimmunoassay using an antiserum raised in rabbits to human PTHrP(1-34) which is specific for the 9-18 region (6).

Results

Effects of exogenous PTHrP(1-34)

Table 1 shows that in all four sheep, administration of PTHrP(1-34) was associated with a significant ($P<0.001$) decrease in the amplitude of the reticular contractions, relative to the preceding control period. Similarly, there were significant reductions in the amplitudes of both the A and B waves of contraction either during the infusion of the PTHrP or during the 15-20 min period following its end or during both periods. The frequency of the B wave of contraction was always decreased during the infusion of PTHrP but that of the reticular contractions was increased, serving to compensate in part for their decreased amplitude of contraction. These results were associated with a loss of rumination and a rise in the plasma concentration of PTHrP(1-34) from 80 to 1840 pmol/l.

Table 1 also shows the comparative stability of the above parameters in two sheep studied under the same conditions but without infusion of peptide. Rumination continued unimpaired during these periods in contrast to its loss during the administration of PTHrP(1-34).

Effects of exogenous PTH(1-34)

Similar results to those shown in Table 1 were obtained in the two sheep infused with PTH(1-34). The mean percentage reductions in amplitude of the reticular, A-wave and B-wave contractions during the period of infusion of the PTH(1-34) were 35, 33 and 28 (one sheep only). In all individual sheep these reductions were significant ($P<0.01$). Also, the frequency of the B-wave contractions was reduced by 58%. Rumination was also inhibited. These results were associated with a rise in plasma PTH(1-34) concentration from 7 to 570 pmol equiv/l.

Induced hypocalcaemia

A 20 ± 3% reduction in plasma Ca^{2+} concentration in each of the four sheep caused similar effects to those associated with the infusion of PTH/PTHrP(1-34), i.e. mean percentage reductions in the amplitude of contraction of the reticulum, A wave and B wave of 25 ± 8 (4), 32 ± 4 and 39 ± 8, respectively. There was also a 29 ± 11% decrease in the frequency of the B waves of contraction and cessation of rumination. There were significant positive linear relationships between the plasma Ca^{2+} concentration and the amplitude of the contractions of the reticulum and of the A wave and also of the frequency of the B-wave contractions. Associated with these results was a doubling of the plasma PTH(1-34) concentration.

To confirm that this comparatively small increase in plasma PTH(1-34) concentration was related to these results, hypocalcaemia was induced in further experiments after pre-treatment with the PTH/PTHrP(1-34) receptor blocker. An 18 ± 2% change in plasma Ca^{2+} concentration then did not cause any significant changes in either the amplitudes or the frequencies of the reticular and B-wave contractions. Only a 24 ± 7% decrease in the A-wave contraction was noted. Rumination persisted in three of the four experiments in which rumination was occurring at the start of the hypocalcaemic period.

Table 1 The effects of the i.v. infusion of PTHrP(1-34) at the rate of 2.1 µg/min, following an injection of 12 µg, of PTHrP(1-34), on reticulo-ruminal contractions in trained, conscious sheep.

Sheep		Period (min)	Reticular contractions		B-wave contractions		A-wave contractions	
			Amplitude (cm)	Frequency (min^{-1})	Amplitude (cm)	Frequency (min^{-1})	Amplitude (cm)	Frequency (min^{-1})
T	Vehicle	18	3.22±0.17 (18)	1.00	3.99±0.26 (14)	0.73	3.87±0.34 (14)	1.07
	PTHrP(1-34)	20	1.92±0.14** (27)	1.23	3.22±0.47 (9)	0.41	3.40±0.19 (29)	0.94
	Vehicle	20	2.09±0.19** (19)	0.95	2.71±0.16* (16)	0.71	2.98±0.18§ (20)	0.88
S	Vehicle	44	4.80±0.16 (46)	1.05				
	PTHrP(1-34)	20	2.64±0.19** (26)	1.24				
	Vehicle	15	1.49±0.11** (18)	1.20				
2	Vehicle	10	7.46±0.09 (8)	0.80	4.38±0.14 (9)	0.83	7.50±0.14 (8)	0.80
	PTHrP(1-34)	20	3.96±0.32** (24)	1.41	4.10±0.31 (10	0.61	5.64±0.26** (24)	1.41
	Vehicle	14	4.04±0.26** (12)	0.86	3.22±0.22* (9	0.64	5.51±0.10** (12)	0.86
U	Vehicle	30	5.68±0.26 (22)	0.73	6.00±0.29 (14)	0.47	4.31±0.14 (25)	0.83
	PTHrP(1-34)	20	2.66±0.24** (22)	1.05	3.30±0.14** (7)	0.33	1.85±0.14** (22)	1.05
	Vehicle	21	1.67±0.18** (12)	0.57	3.85±0.52§ (6)	0.29	1.13±0.26 (14)	0.67
1A	Control	0-19	5.96±0.28 (20)	1.11	6.49±0.19 (52)	0.41	4.82±0.19 (25)	1.11
	Control	68-110	5.86±0.19 (39)	0.94	6.14±0.25 (28)	0.30	4.61±0.30 (33)	0.94
L6	Control	0-42	4.67±1.4 (49)	0.54	2.75±0.37 (15)	0.21	4.42±0.11 (22)	0.70
	Control	104-191	4.93±1.6 (40)	0.49	2.62±0.25 (17)	0.21	4.53±0.18 (38)	0.49

Data are mean±SEM. Figures in parentheses are numbers of measurements. §$P<0.05$ from 1st period of vehicle; *$P<0.01$; **$P<0.001$.

Discussion

The partial inhibition of the cycle of contraction of the reticulo-rumen, coupled with the cessation of rumination observed during the infusion of either PTHrP(1-34) or PTH(1-34) could have been the result of the supra physiological concentrations employed. However, the replication of these results by a mild hypocalcaemic challenge, associated with a small increase in circulating PTH(1-34) suggests that the reduced motility of the reticulo-rumen was probably an effect of the peptide rather than of the hypocalcaemia *per se*. The fact that blockage of the PTH(1-34) receptor largely restores the motility to normal supports this conclusion.

Constipation is a well-known feature of the clinical condition of bovine parturient hypocalcaemia. This may, at least in part, result from the increased concentration of PTH observed, since normalisation of the PTH concentration by injection of a Ca salt leads to the restoration of normal motility of the digestive tract.

Acknowledgements

Financial support from the ISIS fund of the BBSRC is gratefully acknowledged by AD Care. Our thanks are due to Dr SK Abbas for carrying out the hormone assays.

References

1   Cooper CW, Seitz PK, McPherson MB, Selvanayagan P & Rajaraman S (1991) Effects of parathyroid hormonal peptides on the gut. *Contributions to Nephrology* **91** 26-31.
2   Ito M, Ohtsuru A, Enomoto H, Ozeki S-I, Nakashima M, Nakayama T, Shichijo K, Sekine I & Yamashita S (1994) Expression of parathyroid hormone-related peptide in relation to perturbations of gastric motility in the rat. *Endocrinology* **134** 1936-1942.
3   Mok LLS, Ajiwa E, Martin J, Thompson JC & Cooper CW (1989) Parathyroid hormone-related protein relaxes rat gastric smooth muscle and shows cross-desensitisation with parathyroid hormone. *Journal of Bone and Mineral Research* **4** 433-439.
4   Nutt RF, Caulfield MP, Levy JJ, Gibbons SW, Rosenblatt M & McKee RL (1990) Removal of partial agonism from parathyroid hormone (PTH)-related protein-(7-34)NH2 by substitution of PTH amino acids at positions 10 and 11. *Endocrinology* **127** 491-493.
5   Orloff JJ, Wu TL, & Stewart AF (1989) Parathyroid hormone-like proteins: biochemical responses and receptor interactions. *Endocrine Reviews* **10** 476-495.
6   Ratcliffe JG, Norbury S, Stott RA, Heath DA & Ratcliffe JC (1991) Immunoreactivity of plasma-parathyrin-related peptide; three region specific radioimmunoassays and a two-site immunoradiometric assay compared. *Clinical Chemistry* **37** 1781-1787.

The Comparative Endocrinology of Calcium Regulation
Eds C Dacke, J Danks, I Caple & G Flik, pp 177-181
Journal of Endocrinology Ltd, Bristol (1996)

Calcium homeostasis in mole-rats by manipulation of teeth and bone calcium reservoirs

R Buffenstein and T Pitcher

Department of Physiology, University of the Witwatersrand Medical
School, 7 York Road, Parktown 2193, Johannesburg, South Africa
(T Pitcher is now at Department of Biological Sciences, 1.124 Stopford
Building, University of Manchester, Oxford Road, Manchester M13 9PT,
UK)

Mole-rats (*Rodentia*, *Bathyergidae*) live in an extensive maze of underground burrows in the arid and semi-arid regions of the African continent. These chthonic herbivorous rodents have no obvious source of vitamin D. It is, therefore, not surprising that they naturally exist in an impoverished vitamin D state, as indicated by low concentrations of vitamin D metabolites (4), high 1-alpha renal hydroxylase activity (3) and the absence of duodenal vitamin D dependent calbindins (4).

The calciotropic hormones, vitamin D and parathyroid hormone (PTH), play pivotal roles in the regulation of mineral homeostasis. Vitamin D primarily controls intestinal calcium absorption, while PTH regulates renal mineral excretion and manipulates hard tissue reservoirs so that plasma concentrations of calcium are confined to the narrow optimal range vital for normal physiological function (2).

We tested the hypothesis that mineral metabolism, in mole-rats, is not regulated by calciotropic hormonal regulation of intestinal absorption and renal excretion of calcium. We propose that calcium homeostasis is realized by manipulating hard tissue calcium reservoirs. We therefore monitored all aspects of gastro-intestinal absorption and renal excretion of calcium and the calcium content of teeth and bone after vitamin D supplementation and after manipulation of dietary calcium concentration.

Materials and methods

Adult mole-rats (*Cryptomys damarensis*; $n=63$) were housed individually under dim incandescent lighting. Calcium balance was assessed in animals maintained on different diets, namely a mixture of sweet potato and apple (Ca content 31.75 mmol/kg dry matter) or carrot roots (Ca content 66.5 mmol/kg dry matter). In addition some animals received, every three days, a supplement of vitamin D_3 (25 ng/g dry matter food eaten), beginning 6 days prior to experimental monitoring. Food supplied, uneaten food and fecal and urinary output were monitored daily for at least 7 days. Thereafter, animals were anaesthetized and killed by cardiac exsanguination. Plasma, teeth, bone, urinary and fecal samples were frozen at -20 °C for later analyses.

Intestinal calcium uptake was determined using a modified everted gut sac technique (9). Calcium content in samples of food, feces, urine, teeth and bone was determined by atomic absorption spectrophotometry (9). Urinary cAMP was measured using a commercially available kit (TRK432, Amersham, Berks, UK).

This study was approved by The Animal Ethics Committee of the University of the Witwatersrand-Ethics number: 93/68/2b.

Results

Irrespective of dietary calcium content or vitamin D supplementation, mole-rats maintained a constant plasma calcium concentration, and a net intestinal calcium influx. The apparent fractional absorption efficiency (AFA%) for calcium was high and the mode of intestinal calcium absorption appeared passive (Tables 1 and 2).

Renal function, as indicated by urinary calcium output and apparent fractional retention efficiency (AFR%), was constant, regardless of dietary manipulations. Changes in dietary calcium content resulted in an increased net influx of calcium; most of the extra calcium was deposited in teeth and bone (Table 1). Vitamin D supplementation also affected hard tissue calcium content but had no effect on AFA% or AFR% (Table 2). The urinary cAMP concentration after vitamin D supplementation was ten times lower (0.04 ± 0.03 µmol/g creatinine) than in controls (0.43 ± 0.3 µmol/g creatinine), while plasma $1,25(OH)_2D$ concentrations were elevated (Table 2).

Table 1 Calcium balance and dietary calcium content

	Low calcium diet	High calcium diet
Calcium content (mmol/kg)	31.75	66.50
AFA%	91±0.9	95±0.5
AFR%	95±1.1	96±0.4
Tooth calcium content (mmol/g tooth)	4.5±0.1	6.0±0.5[*]
Bone calcium content (mmol/g bone)	4.0±0.2	4.9±0.4[*]
Plasma calcium concentration (mmol/l)	2.7±0.1	2.8±0.1
Plasma $1,25(OH)_2D$ (pg/ml)	14±0.7	12±1.3
^{45}Ca serosal:mucosal ratio	1.1±0.1	1.0±0.1

Data are mean±SEM. *$P<0.05$.

Discussion

Calcium homeostasis appears to be primarily realized by the manipulation of hard tissue - in particular, the calcium content of teeth. All other aspects of calcium handling remained unaffected despite changes in net calcium intake following alterations in either dietary calcium content or vitamin D status. On both the low and high calcium diets a net calcium influx was evident. Irrespective of dietary calcium content, vitamin D status or net calcium retention, plasma calcium concentration was tightly regulated within the normal range for most mammals (8).

Gastrointestinal function

Irrespective of dietary calcium content or vitamin D status, duodenal calcium absorption appeared passive. Lack of active uptake in this region of the gastrointestinal tract, regardless of the vitamin D status, is unusual among mammals (6, 7) and may reflect the absence of vitamin D-dependent duodenal calbindins in mole-rat tissue (4). The predominantly passive intestinal absorption appeared extremely efficient (AFA>90%), indicating that paracellular and endocytic pathways may guarantee efficient extraction of calcium from the gastrointestinal tract. Considering the passive nature of intestinal calcium absorption in this animal, it may not be surprising that absorption was not influenced by vitamin D supplementation.

Table 2 Calcium balance and vitamin D

	D_N	D_S
AFA%	95±0.5	95±0.9
AFR%	96±0.9	98±0.1
Tooth calcium content (mmol/g tooth)	6.0±0.5	8.3±0.9[*]
Bone calcium content (mmol/g bone)	4.9±0.4	5.6±0.4
Plasma calcium concentration (mmol/l)	2.8±0.1	2.7±0.1
Plasma 1,25(OH)$_2$D (pg/ml)	12±1.3	32±5.4[*]
^{45}Ca serosal:mucosal ratio	1.0±0.1	0.98±0.1

Data are mean±SEM. * Indicates $P<0.05$. D_N is the unsupplemented group and D_S is the group receiving an oral vitamin D supplement.

We have shown recently (5) that neither rapid transmembrane Ca^{2+} influx nor slower calcium uptake was stimulated by vitamin D. Given the absence of duodenal calbindins - even in vitamin D supplemented animals - it appears that vitamin D-mediated, genomic and non-genomic, actions on intestinal calcium uptake are absent in mole-rats (5). Gastrointestinal absorption in these underground animals differs from that of most mammals (7) and is independent of vitamin D.

Renal function

The increase in dietary calcium intake resulted in an increase in the amount of calcium absorbed. Surprisingly, this was not associated with any increase in urinary output. Regardless of dietary calcium content or vitamin D status, most of the calcium, present in the glomerular filtrate is efficiently extracted, resulting in AFR% approaching the physiological maximum (>95%). The urinary cAMP concentration, assumed to reflect PTH activity, declined when animals were given vitamin D supplements, suggesting that vitamin D regulation of PTH in mole-rats does not differ in any way from that previously reported for other mammals (8). The presumed down-regulation of PTH, however, did not appear to affect renal handling of calcium. A high AFR%, irrespective of dietary calcium content and calciotropic hormone concentration, implies that mole-rats cannot effectively control renal passive calcium absorption. This appears to be unresponsive to endocrine control.

Calcium homeostasis, therefore, does not appear to be regulated at the level of the kidney and is independent of dietary calcium content or vitamin D status, but involves manipulation of bony compartments that act as internal calcium reservoirs to tightly regulate plasma calcium concentration.

Hard tissues

Calcium content of mole-rat incisors varied considerably with diet and vitamin D supplementation; the fluctuation in bone calcium content was considerably smaller. These two calcium pools may thus assist in calcium homeostasis. Calcium stored in bone may act as a reservoir and be drawn upon when calcium homeostasis is threatened. Calcium stored in teeth, however, cannot be reabsorbed and rather is lost when these evergrowing incisors are worn down. Mole-rats dig with these chisel-like teeth and also spend considerable periods grinding their teeth when resting. The teeth most likely serve as a sink for the excess retained mineral. Vitamin D supplementation, also affected tooth mineral content, although bone mineral content was unchanged. Similar selective dumping of calcium in teeth and not bone has been reported in rats (1).

Conclusion

Calcium homeostasis in subterranean mole-rats does not appear to be regulated at the level of the kidney or intestine. Only bone and teeth mineral content changed in response to dietary calcium manipulation and vitamin D supplementation. Rather, calcium homeostasis is facilitated by manipulating hard tissue calcium stores.

Acknowledgements

We gratefully acknowledge assistance from the MRC Mineral Metabolism Research Unit, in particular from Professor John Pettifor, Mr GP Moodley, and Ms D Zachen and funding from the MRC (South Africa).

References

1   Bauer GCH (1954) Rate of bone formation in healing fracture determined in rats by means of radiocalcium. *Acta Orthopaedica Scandinavica* **23** 169-191.

2   Broadus AE (1993) Physiological function of calcium, magnesium and phosphorus and mineral ion balance. *Primer on the Metabolic Bone Diseases and Disorders of Mineral Metabolism*, pp 41-46. Ed MJ Favus. New York: Raven Press.

3   Buffenstein R, Sergeev IN & Pettifor JM (1993) Vitamin D hydroxylases and their regulation in a naturally vitamin D deficient subterranean mammal, the naked mole-rat (*Heterocephalus glaber*). *Journal of Endocrinology* **138** 59-64.

4   Buffenstein R, Jarvis JUM, Opperman LA, Cavaleros M, Ross FP & Pettifor JM (1994) Subterranean mole-rats naturally have an impoverished calcitriol status, yet synthesise calcitriol metabolites and calbindins. *European Journal of Endocrinology* **130** 402-409.

5   Buffenstein R, Sergeev IN & Pettifor JM (1994) Absence of calcitriol-mediated
     nongenomic actions in isolated intestinal cells of the Damara mole-rat (*Cryptomys
     damarensis*). *General and Comparative Endocrinology* **95** 25-29.

6   Favus MJ (1985) Factors that influence absorption and secretion of calcium in the small
     intestine and colon. *American Journal of Physiology* **248** G147-G157.

7   Norman AW (1990) Intestinal calcium absorption: a vitamin D-hormone mediated
     adaptive response. *American Journal of Clinical Nutrition* **51** 290-300.

8   Norman AW & Litwack G (1987) The calcium regulating hormones: vitamin D,
     parathyroid and calcitonin. *Hormones*, pp 356-396. Ed AW Norman. New York:
     Academic Press.

9   Pitcher T, Buffenstein R, Keegan JD, Moodley GP & Yahav S (1992) Dietary calcium
     content, calcium balance and mode of uptake in a subterranean mammal, the Damara
     mole-rat. *Journal of Nutrition* **122** 108-114.

The Comparative Endocrinology of Calcium Regulation
Eds C Dacke, J Danks, I Caple & G Flik, pp 183-189
Journal of Endocrinology Ltd, Bristol (1996)

Current status of the pathogenesis of humoral hypercalcemia of malignancy in dogs

T J Rosol, L A Nagode, D J Chew[1], C L Steinmeyer
and C C Capen
Department of Veterinary Biosciences, The Ohio State University, 1925
Coffey Road, Columbus, OH 43210 USA and [1]Department of Veterinary
Clinical Sciences, The Ohio State University, 601 Vernon Tharp Street,
Columbus, OH 43210 USA

Humoral hypercalcemia of malignancy in dogs

Humoral hypercalcemia of malignancy (HHM) is a common clinical syndrome in dogs (3, 14). It occurs most frequently with lymphoma or adenocarcinomas derived from the apocrine glands of the anal sac (11, 21). HHM also occurs sporadically in dogs with various tumors including squamous cell carcinomas, carcinomas of the lung, pancreas, thyroid, liver, mammary gland, or nasal cavity, and multiple myeloma (3, 14). HHM is much less common in other domestic animals, but has been reported in cats and horses (14, 18). Hypercalcemia in dogs with HHM is due principally to increased osteoclastic bone resorption and renal reabsorption of calcium (12-14)

Clinical signs of HHM in dogs

Dogs with HHM develop increased serum concentrations of total and ionized calcium and hypophosphatemia (3, 11). The hypercalcemia induces polyuria and secondary polydipsia due to impairment of renal concentrating mechanisms (3). Dogs also develop muscle weakness, depression, dehydration, anorexia and vomiting (3). Clinical signs are variable and are most severe in dogs with rapid increases of serum ionized calcium or dehydration. Hypercalcemia can lead to life-threatening renal failure due to nephrosis and renal mineralization (2).

Lymphoma and HHM

Lymphoma is the most common cause of HHM in dogs (3). Approximately 20% of dogs with multicentric or thymic lymphoma develop HHM (6). There are two potential mechanisms for the development of hypercalcemia: (i) secretion of humoral factor(s) by lymphoma cells with distant effects in bone and kidney, and (ii) local bone resorption in the marrow secondary to tumor invasion (10, 21). Most dogs with hypercalcemia and lymphoma have the humoral form. Dogs with bone metastases may have a combination of HHM and local bone resorption associated with intramedullary lymphoma. Humoral factors that may be associated with the pathogenesis of HHM in dogs with lymphoma include parathyroid hormone-related protein (PTHrP), interleukin-1 and tumor necrosis factors (14, 17). Most lymphomas associated with HHM in dogs are derived from T-cells (6, 21). This suggests that HHM associated with lymphoma in dogs may be a valuable animal model of HHM associated with human T-cell lymphotrophic virus (HTLV)-1-induced lymphoma in human beings (4). Unfortunately, cell lines or tumor lines of canine lymphoma propagated in nude mice have not been developed since lymphoma cells from dogs are fastidious and resistant to growth *in vitro* or *ex vivo*. This has slowed progress on the pathogenesis of HHM associated with this form of T-cell lymphoma in dogs.

Adenocarcinomas derived from apocrine glands of the anal sac and HHM

Adenocarcinomas derived from apocrine glands of the anal sac are an uncommon tumors that occur in older dogs with a higher prevalence in females (11). The tumors are slow growing, but have a high incidence of metastasis to internal lymph nodes and viscera. Apocrine adenocarcinomas of the anal sac are associated with a high incidence of HHM which occurs in approximately 30-50% of affected dogs (3, 11). Surgical removal of the tumor results in temporary remission from the clinical signs associated with HHM. Remission persists until development of intra-abdominal metastases. PTHrP is the primary humoral factor associated with HHM and apocrine adenocarcinomas of the anal sac (16, 17).

Nude mouse model (CAC-8) of HHM and apocrine adenocarcinoma of the anal sac

A serially transplantable tumor line derived from a canine apocrine adenocarcinoma of the anal sac has been developed in nude mice (15). The tumor line (CAC-8) has been propagated for over 30 passages. Nude mice with transplanted CAC-8 develop clinical signs of HHM similar to dogs and humans, including hypercalcemia, hypophosphatemia and hypercalciuria (15). The hypercalcemia is dependent on osteoclastic bone resorption, renal reabsorption of calcium and calcium absorption from the intestinal tract (13).

The nude mouse model of HHM differs in two aspects compared with the pathogenesis of HHM in dogs and humans. Nude mice with HHM have increased serum concentrations of 1,25-dihydroxyvitamin D and increased osteoblastic bone formation (15). Most rodent models of HHM are associated with increased serum 1,25-dihydroxyvitamin D concentrations (14). This may be due to an increased sensitivity of rodent kidney to hypophosphatemia which results in stimulation of renal 1-α-hydroxylase and synthesis of 1,25-dihydroxyvitamin D. The increased osteoblastic bone formation that is present in nude mice with CAC-8 is not able to maintain bone mass and there is a progressive loss of trabecular and cortical bone. This is interpreted to represent a relative uncoupling of bone resorption and formation. However, bone formation is increased in tumor-bearing nude mice and this contrasts with the presence of decreased bone formation in humans with HHM (14). The cause of decreased bone formation in humans is unknown but it may be associated with tumor-induced cachexia or tumor-related cytokine production (1).

The CAC-8 tumor line produces the three biologic activities commonly associated with tumors that induce HHM, including PTH-like activity, transforming growth factor (TGF)α, and TGFβ (8). The PTH-like activity and the majority of the bone resorbing activity in the tumor is due to the presence of PTHrP (8, 16). The roles of TGFα and TGFβ are not known. They may function in combination with PTHrP to induce osteoclastic bone resorption. Alternatively, TGFβ may stimulate the production of PTHrP by the tumor cells, since TGFβ has been reported to induce increased expression of PTHrP in multiple cell types, including squamous carcinomas (9).

PTHrP, PTH and 1,25-dihydroxyvitamin D in dogs with HHM

PTHrP in dogs with HHM

N-terminal radioimmunoassay (Incstar, Stillwater, MN) (17) PTHrP was undetectable (<1.8 pM) in aprotinin-treated plasma from normal dogs. Dogs with hypercalcemia (total serum Ca>12 mg/dl) and apocrine adenocarcinomas of the anal sac had markedly increased PTHrP concentrations with a range of 14-99 pM. Dogs with normocalcemia and anal sac adenocarcinomas had undetectable PTHrP concentrations or PTHrP levels less than 7.6 pM. Dogs with anal sac adenocarcinomas and hypercalcemia had a strong linear correlation between PTHrP and the total serum calcium concentration. This suggests that PTHrP is the principal hypercalcemic factor in this form of HHM. Most dogs with hypercalcemia and lymphoma had detectable PTHrP concentrations (1.8-16.7 pM) and most dogs with normocalcemia and lymphoma had undetectable PTHrP concentrations. In contrast to dogs with apocrine adenocarcinomas of the anal sac, dogs with lymphoma and hypercalcemia did not have a significant correlation between PTHrP and serum calcium concentrations. This indicates that PTHrP may be a valuable serum marker of HHM in dogs with lymphoma; however, PTHrP likely functions synergistically with other tumor-related factors (e.g. cytokines) to induce hypercalcemia (14). Dogs with hypercalcemia and miscellaneous tumors had detectable PTHrP levels that ranged from 3.8-38.4 pM. Six dogs with parathyroid adenomas had undetectable levels of PTHrP. Most dogs with normocalcemia and miscellaneous tumors had undetectable levels of PTHrP or concentrations less than 3.5 pM. Plasma PTHrP concentrations decreased significantly in dogs with HHM and anal sac adenocarcinomas or lymphoma after treatment (surgical removal or intraoperative radiation in dogs with adenocarcinomas and combined chemotherapy in dogs with lymphoma) and return to normocalcemia.

Two-site immunoradiometric assay for PTHrP (20) The mean PTHrP concentration in 19 normal dogs measured by a two-site IRMA was 0.9 pM. Dogs with HHM had plasma PTHrP concentrations which ranged from 2-60 pM. The mean plasma PTHrP concentration in dogs with HHM was 12-fold greater than in normal dogs.

PTH in dogs with HHM

N-terminal PTH radioimmunoassay (INS human PTH Assay, Nichols Institute, San Juan Capistrano, CA) (17) Serum PTH concentrations were usually in the normal range (12-34 pg/ml) for most dogs with HHM. Hypercalcemic dogs with parathyroid adenomas had increased serum PTH concentrations (mean=83±38 pg/ml). Serum PTH concentrations were measured in hypercalcemic dogs with anal sac adenocarcinomas or lymphoma before and after surgery, radiation or chemotherapy. Post-treatment samples were collected after initiation of therapy when the serum calcium concentration was in the normal range (<12.0 mg/dl). Correction of hypercalcemia in dogs with HHM was associated with a marked increase in serum PTH concentrations from 18±7 to 109±78 pg/ml. These findings were interpreted to indicate that PTH secretion was suppressed in dogs with HHM, which was rapidly reversed after return to normocalcemia with a rebound of serum PTH to supranormal levels.

1,25-dihydroxyvitamin D in dogs with HHM

Serum 1,25-dihydroxyvitamin D was usually normal (20-50 pg/ml) or suppressed in dogs with HHM (17). There were distinct subgroups of dogs with hypercalcemia and lymphoma or apocrine adenocarcinomas of the anal sac that had increased serum concentrations of 1,25-dihydroxyvitamin D. Evaluation of serum 1,25-dihydroxyvitamin D before and after treatment of dogs with HHM revealed reduced serum 1,25-dihydroxyvitamin D concentrations (16±10 compared with 38±20 pg/ml) after therapy and return to normocalcemia. Serum 1,25-dihydroxyvitamin D concentration did not have a significant linear correlation with serum calcium, PTH, or PTHrP in dogs with HHM. These data suggest that increased serum 1,25-dihydroxyvitamin D concentrations may contribute to the development of hypercalcemia in some dogs with HHM. The source of increased serum 1,25-dihydroxyvitamin D in dogs with HHM is uncertain but it may represent renal or tumor production. The serum 1,25-dihydroxyvitamin D concentrations may be variable in most dogs with HHM due to the interactions of multiple stimulatory (hypophosphatemia, increased PTHrP) and inhibitory (hypercalcemia, low PTH concentrations) influences on renal 1,25-dihydroxyvitamin D synthesis.

Canine PTHrP cDNA cloning and sequencing (Genbank accession #U15593)

A lambda phage cDNA library was prepared from poly(A⁺)RNA isolated from a canine apocrine adenocarcinoma of the anal sac (CAC-8) associated with HHM (19). An 1166-base pair (bp) complementary DNA (CP3) was cloned which encoded for canine PTHrP (cPTHrP). The cDNA (CP3) was sequenced in both directions. Base pairs 1-145 and 146-167 of CP3 encode for the 5'-untranslated region (UTR) and are 74% and 77% homologous to the 5'-UTRs of exons 1A and 3 of human PTHrP (hPTHrP) respectively (7). The cPTHrP cDNA is unique since it has homology to exon 1A of hPTHrP. This suggests that dogs utilize a promoter similar to P1 of hPTHrP which has not been demonstrated in other species (5). The coding region of CP3 (bp 168-698) has an open reading frame of 177 amino acids. The coding region of CP3 is 91, 82, 83, and 72% homologous to the coding regions of human, rat, mouse and chicken PTHrP respectively. The 3'-UTR (bp 702-1166) of CP3 is 96, 88 and 87% homologous to the terminal exon of human, rat and mouse PTHrP respectively. The predicted mature protein contained 141 amino acids and had high homology (99%) to rodent and human PTHrP in the first 111 amino acids (Fig. 1). The C-terminal region (amino acids 112-141) had a low degree of homology (40 and 67%) to rodent and human PTHrP. The sequence of canine PTHrP will facilitate further investigations into the role of PTHrP in the pathogenesis of HHM in dogs. Immunologic assays for hPTHrP that utilize antibodies to the N-terminal region or mid-region of PTHrP should be useful for measuring circulating PTHrP in dogs due to the high degree of homology between these two species.

```
                .          .          .          .          .
cPTHrP  AVSEHQLLHDKGKSIQDLRRRFFLHHLIAEIHTAEIRATSEVSPNSKPAP  ( 50)
hPTHrP  ...............................................S.  ( 50)
rPTHrP  .................................................  ( 50)
mPTHrP  .................................................  ( 50)


cPTHrP  NTKNHPVRFGSDDEGRYLTQETNKVETYKEQPLKTPGKKKKGKPGKRKEQ  (100)
hPTHrP  ..................................................  (100)
rPTHrP  ...............................................R..  (100)
mPTHrP  ...............................................R..  (100)


cPTHrP  EKKKRRTRSAWLNSGVAESGLEGDHPYDISATS--LELNLRRH        (141)
hPTHrP  ............D...TG.......LS.T.T..--R..DS...        (141)
rPTHrP  ...........--P.TTG...LE.PQPHT.P..TS..PSS.T.        (141)
mPTHrP  ...........--PST.A...LE.PLPHT.R..--..PS..T.        (139)
```

Fig. 1 Canine (c), human (h), rat (r) and mouse (m) PTHrP (1-139 or 141)

Summary

Humoral hypercalcemia of malignancy in dogs represents an important syndrome from two aspects: (1) it is a significant clinical problem in veterinary medicine, and (2) it is a useful spontaneous animal model of cancer-associated hypercalcemia. There are few spontaneous tumors in domestic or laboratory animals that are consistently associated with HHM. T-cell lymphoma and apocrine adenocarcinomas of the anal sac in dogs are examples of spontaneous HHM in animals. These tumors represent two pathogenic mechanisms of HHM. Apocrine adenocarcinomas of the anal sac induce hypercalcemia principally by secreting biologically active PTHrP into the circulation in an unregulated manner. In contrast, hypercalcemia associated with T-cell lymphoma in dogs has a more complex pathogenesis. Hypercalcemia in dogs with lymphoma is likely induced by the synergistic action of multiple tumor-related humoral factors including PTHrP, cytokines and 1,25-dihydroxyvitamin D.

References

1   Barengolts EI, Lathon PV, Lindh F & Kukreja S C (1994) Cytokines may be responsible for the inhibited bone formation in hypercalcemia of malignancy. *Journal of Bone and Mineral Research* **9** S139.

2   Chew DJ & Capen CC (1980) Hypercalcemic nephropathy and associated disorders. *Current Veterinary Therapy VII*, pp 1067-1072. Ed R W Kirk. Philadelphia, PA: Saunders.

3   Chew DJ, Nagode LA & Carothers M (1992) Disorders of calcium: Hypercalcemia and hypocalcemia. *Fluid Therapy in Small Animal Practice*, pp 116-176. Ed SP DiBartola. Philadelphia, PA: Saunders.

4   Fukumoto S, Matsumoto T, Ikeda K, Yamashita T, Watanabe T, Yamaguchi K, Kiyokawa T, Takatsuki K, Shibuya N & Ogata E (1988) Clinical evaluation of calcium metabolism in adult T-cell leukemia/lymphoma. *Archives of Internal Medicine* **148** 921-925.

5   Gillespie MT & Martin TJ (1994) The parathyroid hormone-related protein gene and its expression. *Molecular and Cellular Endocrinology* **100** 143-147.

6   Greenlee PG, Filippa DA, Quimby FW, Patnaik AK, Calvano SE, Matus RE, Kimmel M, Hurvitz AI & Lieberman PH (1990) Lymphomas in dogs: a morphologic, immunologic and clinical study. *Cancer* **66** 480-490.

7   Mangin M, Ikeda K, Dreyer BE & Broadus AE (1990) Identification of an up-stream promoter of the human parathyroid hormone-related peptide gene. *Molecular Endocrinology* **4** 851-858.

8   Merryman JI, Rosol TJ, Brooks CL & Capen CC (1989) Separation of parathyroid hormone-like activity from transforming growth factor-α and -β in the canine adenocarcinoma (CAC-8) model of humoral hypercalcemia of malignancy. *Endocrinology* **124** 2456-2563.

9   Merryman JI, DeWille JW, Werkmeister JR, Capen CC & Rosol TJ (1994) Effects of transforming growth factor-β on parathyroid hormone-related protein production and ribonucleic acid expression by a squamous carcinoma cell line *in vitro*. *Endocrinology* **134** 2424-2430.

10   Meuten DJ, Kociba GJ, Capen CC, Chew DJ, Segre GV, Levine L, Tashjian AHJ, Voelkel EF & Nagode LA (1983) Hypercalcemia in dogs with lymphosarcoma: biochemical, ultrastructural, and histomorphometric investigations. *Laboratory Investigations* **49** 553-562.

11   Meuten DJ, Segre GV, Capen CC, Kociba GJ, Voelkel EF, Levine L, Tashjian AH, Chew DJ & Nagode LA (1983) Hypercalcemia in dogs with adenocarcinoma derived from apocrine glands of the anal sac: biochemical and histomorphometric investigations. *Laboratory Investigations* **48** 428-435.

12   Norrdin RW & Powers BE (1983) Bone changes in hypercalcemia of malignancy in dogs. *Journal of the American Veterinary Medical Association* **183** 441-444.

13   Rosol TJ & Capen CC (1987) The effect of low calcium diet, mithramycin, and dichlorodimethylene bisphosphonate on humoral hypercalcemia of malignancy in nude mice transplanted with the canine adenocarcinoma tumor line (CAC-8). *Journal of Bone and Mineral Research* **2** 395-405.

14   Rosol TJ & Capen CC (1992) Biology of disease: mechanisms of cancer-induced hypercalcemia. *Laboratory Investigations* **67** 680-702.

15   Rosol TJ, Capen CC, Weisbrode SE & Horst RL (1986) Humoral hypercalcemia of malignancy in nude mouse model of a canine adenocarcinoma derived from apocrine glands of the anal sac. *Laboratory Investigations* **54** 679-688.

16   Rosol TJ, Capen CC, Danks JA, Suva LJ, Steinmeyer CL, Hayman J, Ebeling PR & Martin TJ (1990) Identification of parathyroid hormone-related protein in canine apocrine adenocarcinoma of the anal sac. *Veterinary Pathology* **27** 89-95.

17   Rosol TJ, Nagode LA, Couto CG, Hammer AS, Chew DJ, Peterson JL, Ayl RD, Steinmeyer CL & Capen CC (1992) Parathyroid hormone (PTH)-related protein, PTH, and 1,25-dihydroxyvitamin D in dogs with cancer-associated hypercalcemia. *Endocrinology* **131** 1157-1164.

18   Rosol TJ, Nagode LA, Robertson JT, Leeth BD, Steinmeyer CL & Allen CM (1994) Humoral hypercalcemia of malignancy associated with ameloblastoma in a horse. *Journal of the American Veterinary Medical Association* **204** 1930-1933.

19   Rosol TJ, Steinmeyer CL, McCauley LK, Gröne A, DeWille JW & Capen CC (1995) Sequences of the cDNAs encoding canine parathyroid hormone-related protein and parathyroid hormone. *Gene* **60** 241-243.

20   Weir EC (1992) Hypercalcemia and malignancy. *Proceedings of the 10th American College of Veterinary Internal Medicine Forum* **10** 640-612.

21   Weir EC, Norrdin RW, Matus RE, Brooks MB, Broadus AE, Mitnick M, Johnston SD & Insogna KL (1988) Humoral hypercalcemia of malignancy in canine lymphosarcoma. *Endocrinology* **122** 602-608.

The Comparative Endocrinology of Calcium Regulation
Eds C Dacke, J Danks, I Caple & G Flik, pp 191-196
Journal of Endocrinology Ltd, Bristol (1996)

Comparative analysis of the bovine and human calcium-sensing receptors

S H S Pearce and R V Thakker

Medical Research Council Molecular Endocrinology Group, Hammersmith
Hospital, DuCane Road, London W12 0NN, UK
(S H S Pearce is now at Department of Medicine, The Medical School,
Framlington Place, Newcastle upon Tyne NE2 4HH, UK)

Introduction

The gene encoding a 120 kD calcium-sensing receptor (CaR) has recently been
identified by expression cloning from bovine parathyroid tissue (2) and has been
designated the symbol BoPCaR1. The CaR receptor has a large extracellular
ligand-binding domain and the seven hydrophobic transmembrane domains typical of
a G protein-coupled receptor (2). The CaR gene consists of seven exons which span
45 kb of genomic DNA (3). The 612 amino acid extracellular N-terminal domain is
encoded by exons 2 to the 5' portion of exon 7, while the transmembrane domains and
the C-terminal tail are encoded by the remainder of exon 7. The human homologue of
this calcium-sensing receptor gene, which is referred to as PCaR1, is located on
chromosome 3q21-q24 (11) and its role in the aetiology of familial benign
(hypocalciuric) hypercalcaemia (FBH) has been investigated. Heterozygosity for
PCaR1 mutations has been established in FBH kindreds (6, 8, 10, 11) and
homozygosity for PCaR1 mutation in two FBH families with children affected by
neonatal severe hyperparthyroidism has been reported (8, 11). We report the coding
DNA sequence (EMBL/Genbank accession number X81086) and amino acid sequence
of the human PCaR1 together with a comparative analysis between the human and the
bovine CaR sequences and a phylogenetic comparison between the CaRs and other G
protein-coupled receptors.

Methods

Genomic DNA was extracted from peripheral-blood leukocytes of 15 unrelated
individuals and the six coding exons of the PCaR1 gene and nine of the 12 intron-exon
splice boundaries were amplified by the polymerase chain reaction (PCR) using 12
pairs of oligonucleotide primers (10, 11). PCR amplification reactions were carried out
in a final volume of 50 μl with 100 ng DNA, 50 pmol of each primer, 200 μM dNTPs,
1 mM magnesium chloride, 50 mM potassium chloride, 10 mM tris HCl (pH 8.3) and
1 U Taq DNA polymerase (Gibco-BRL, Paisley, Strathclyde, UK). An initial
denaturation at 95 °C for 5 min was used followed by 40 cycles, each for 30 s of 94 °C
for denaturation, 65 °C for annealing and 72 °C for extension of template DNA. The

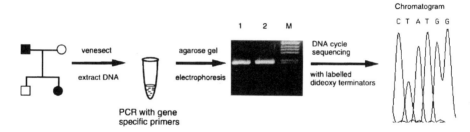

Fig. 1 Schematic representation of cycle sequencing with fluorochrome-labelled dideoxy terminators from a PCR template. Genomic DNA is extracted from peripheral blood leukocytes and used with 12 pairs of PCaR1 sequence-specific oligonucleotide primers to amplify a region of the PCaR1 gene by PCR. The PCR product is then electrophoresed on a 2% agarose gel and gel purified by spinning through glass wool. The size-selected purified DNA is then used as a template for the cycle sequencing reaction which incorporates fluorochrome-labelled dideoxy terminators into the product. The DNA sequence of both strands can then be resolved as the colour peaks of the chromatogram by a laser detection system.

PCR products were then gel purified and both strands were analysed (Fig. 1) by Taq polymerase cycle sequencing with fluorochrome-labelled dideoxy terminators resolved on a semi-automated detection system (Applied Biosystems 373A, Warrington, Cheshire, UK). Comparative sequence analysis of the human PCaR1 sequence and the BoPCaR1 sequence was performed using the FASTA and BESTFIT software (University of Wisconsin Genetics Group). For phylogenetic analysis, the amino acid sequences of members of various sub-groups were retreived from Genbank and used with the published BoPCaR1 amino acid sequence and our deduced human PCaR1 amino acid sequence in the PILEUP program of the University of Wisconsin Genetics Group.

Results

Comparison of human and bovine sequences

The sequence of the human calcium-sensing receptor (PCaR1) was found to have 93% amino acid identity with the BoPCaR1 (2) over its entire length (see Fig. 2). However, the human PCaR1 is seven amino acids shorter than the bovine protein, having 1078 amino acids compared with 1085 in the BoPCaR1 (2). Thus, the human PCaR1 lacks a serine residue (Ser14) in the N-terminal signal peptide sequence as well as six residues (Gln928, Gln929, Ser943, Pro944, His945, Asn946) in the intracellular C-terminal tail. The transmembrane domains, extracellular and cytoplasmic loops are the most highly conserved regions between the bovine and human CaR sequence with 98% amino acid identity between codons 613 and 862. There is also a striking 96% conservation of amino acid identity in the extracellular domain (codons 1 to 612) between the human and bovine CaRs. In the cytoplasmic tail of the human and bovine receptors there is a greater sequence divergence with 82% amino acid identity between codon 863 and the

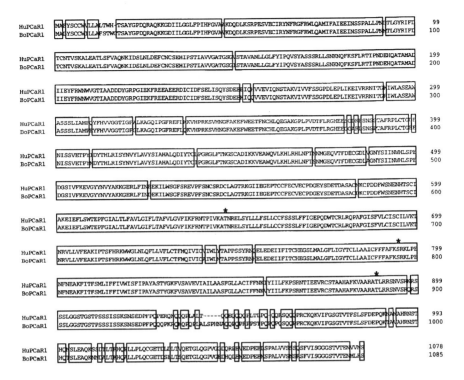

Fig. 2 Amino acid alignment of human and bovine calcium receptors. The amino acid sequence of the human PCaR1 (HuPCaR1) was deduced from the DNA sequence and aligned with the published BoPCaR1 amino acid sequence (2) using the FASTA program. Amino acids are numbered in the right-hand margin to correspond to the position in the human sequence. Identical amino acid residues are highlighted by the boxed areas. Consensus protein kinase C phosphorylation sites that are conserved between human and bovine CaRs are shown by asterisks. The DNA and amino acid sequence of the human receptor have been deposited with EMBL/Genbank under the accession number X81086.

C-terminal. Although the C-terminals of the human and bovine receptors are the most divergent domains they are both serine and threonine rich and thus the human CaR has many potential protein kinase A phosphorylation sites (9). In addition, the intracellular loops and cytoplasmic tail of the BoPCaR1 is reported to have four potential protein kinase C phosphorylation sites (2) but the most C-terminal of these has a weak consensus sequence (9) and our analysis reveals that this is not conserved within the human CaR.

Phylogenetic analysis of calcium receptors and other G protein-coupled receptors

The calcium receptors are not closely related to other G protein-coupled receptors but do share limited homology with the sub-group of metabotropic glutamate receptors

193

(MGluR) (2). This sub-group of MGluRs is itself quite diverse, with the PCaRs having most homology with the type 1 and type 5 MGluRs. Overall the human PCaR1 has 30% amino acid identity with the rat type 1 MGluR, but there are several short stretches of greater homology in the N-terminus with up to 50% similarity in the hydrophobic 'A' segment (2). In the extracellular domains, the rat MGluR1 has 29% amino acid identity with the human PCaR1. Over the transmembrane domains, extracellular and intracellular loops there is 30% amino acid identity between the human PCaR1 and rat MGluR1 but this identity falls to only 26% over the intracellular C-terminal tail. The phylogenetic analysis of the human and bovine PCaRs with the family of rat MGluRs and various other G protein-coupled receptors is shown in Fig. 3. The sub-groups of CaRs and MGluRs appear from this homology analysis to be evolutionarily distinct from the glycoprotein hormone, adrenergic, muscarinic, melanocortin and PTH families of receptors.

Discussion

Our comparative analysis has revealed the high degree of sequence conservation between human and bovine calcium receptors which are markedly divergent from other members of the super-family of G protein-coupled receptors. This high degree of structural conservation between the human PCaR1 and the BoPCaR1 reflect the important function of the receptor in the maintenance of extracellular calcium and magnesium concentrations which are tightly regulated in all mammals (1). In particular the extracellular ligand-binding domains and the signal transducing membrane-spanning domains show remarkable conservation between the human and bovine receptors which must reflect the common signal transduction pathways in both species as well as the common ligands (2). The most divergent area between the human PCaR1 and the BoPCaR1 is the intracellular C-terminal tail. Phosphorylation of this intracellular tail is a necessary step in the rapid down-regulation of some G protein-coupled receptors such as the β-adrenergic receptor (4, 5) and it has been demonstrated that phosphorylation by both specific (β-adrenergic receptor kinase) and non-specific kinases may be a prerequisite of arrestin binding to the cytoplasmic domains which is an important regulator of receptor activity (4, 5, 7). The numerous conserved protein kinase A and the three conserved protein kinase C phosphorylation sites (2, 9) within the C-terminus of the human PCaR1 and BoPCaR1 indicated that there is likely to be some functional redundancy in the pathways leading to the rapid down-regulation of the CaRs. The phylogenetic analysis we have carried out based on a comparison of amino acid homology with other G protein-coupled receptors indicates that the human and bovine CaRs are most closely related to the large sub-group of metabotropic glutamate receptors which also share a large extracellular ligand-binding domain. However, the CaRs have little homology to the glycoprotein hormone sub-group of receptors which also have large extracellular domains. These comparative analysis studies will help in defining the important functional role of the calcium-sensing receptors in the maintenance of extracellular homeostasis and the elucidation of the human PCaR1 sequence will facilitate mutational analysis studies of FBH and the disorders of calcium homeostasis (12).

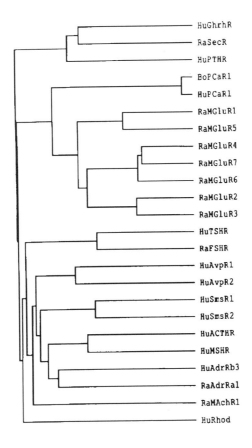

Fig. 3 Phylogenetic alignment of human and bovine calcium receptors with other G protein-coupled receptors. The amino acid sequences of the human and bovine calcium receptors were compared for homology to those of other G protein-coupled receptors using the PILEUP program of the University of Wisconsin Genetics Group. The human PCaR1 and BoPCaR1 calcium receptors were most homologous to the family of metabotropic glutamate receptors particularly MGluR1 and MGluR5, here represented as the rat amino acid sequences (RaMGluR1-7). The comparisons between these and the human TSH-receptor (HuTSHR), rat FSH-receptor (RaFSHR), human ACTH-receptor (HuACTHR), human MSH-receptor (HuMSHR), human β3 adrenergic-receptor (HuAdrRb3), rat α1 adrenergic-receptor (RaAdrRa1), rat type 1 muscarinic cholinergic receptor (RaMAchR1), human rhodopsin (HuRhod), human PTH-receptor (HuPTHR), human growth hormone releasing hormone receptor (HuGhrhR), rat secretin receptor (RaSecR), human type 1 and 2 vasopressin receptors (HuAvpR 1+2) and human type 1 and 2 somatostatin receptors (HuSmsR 1+2) are also illustrated.

Acknowledgements

We are grateful to the Medical Research Council (UK) for support, and to Drs A Soutar and J Morrison for helpful discussions. SHSP is an MRC training fellow.

References

1   Brown EM (1991) Extracellular Ca^{2+} sensing, regulation of parathyroid cell function, and role of Ca^{2+} and other ions as extracellular (first) messengers. *Physiological Reviews* **71** 371-411.

2   Brown EM, Gamba G, Riccardi D, Lombardi M, Butters B, Kifor O, Sun A, Hediger MA, Lytton J & Hebert SC (1993) Cloning and characterization of an extracellular Ca^{2+}-sensing receptor from bovine parathyroid. *Nature* **366** 575-580.

3   Capuano IV, Krapcho KJ, Hung BC, Brown EM, Hebert SC & Garrett JF (1994) Characterisation of the human calcium receptor gene. *Journal of Bone and Mineral Research* **9** S145.

4   Collins S, Lohse MJ, O'Dowd B, Caron MG & Lefkowitz RJ (1991) Structure and regulation of G protein-coupled receptors: the β2-adrenergic receptor as a model. *Vitamins and Hormones* **46** 1-39.

5   Hausdorff WP, Caron MG & Lefkowitz RJ (1990) Turning off the signal: desensitization of β-adrenergic receptor function. *FASEB Journal* **4** 2881-2889.

6   Heath HIII, Odelberg S, Brown D, Hill VM, Robertson M, Jackson CE, Teh BT, Hayward N, Larsson C, Buist N, Garrett J & Leppert MF (1994) Sequence analysis of the parathyroid cell calcium receptor (CaR) gene in familial benign hypercalcemia (FBH): a multiplicity of mutations. *Journal of Bone and Mineral Research* **9** S414.

7   Inglese J, Freedman NJ, Koch WJ & Lefkowitz RJ (1993) Structure and mechanism of the G protein-coupled receptor kinases. *Journal of Biological Chemistry* **268** 23 735-23 738.

8   Janicic N, Pausova Z, Cole DEC & Hendy GN (1995) Insertion of an Alu sequence in the human Ca2+ sensing receptor gene in familial hypocalciuric hypercalcemia and neonatal severe hyperparathyroidism. *American Journal of Human Genetics* **56** 880-886.

9   Kemp BE & Pearson RB (1991) Protein kinases and phosphorylation site sequences. *Methods in Enzymology* **200** 62-77.

10   Pearce SHS, Trump D, Wooding C, Besser GM, Chew SL, Grant DB, Heath DA, Hughes IA, Paterson CR, Whyte MP & Thakker RV (1995) Calcium-sensing receptor mutations in familial benign hypercalcaemia and neonatal hyperparathyroidism. *Journal of Clinical Investigation* **96** 2683-2692.

11   Pollak MR, Brown EM, Chou YWH, Hebert SC, Marx SJ, Steinmann B, Levi T, Seidman CE & Seidman JG (1993) Mutations in the human Ca^{2+}-sensing receptor gene cause familial hypocalciuric hypercalcaemia and neonatal severe hyperparathyroidism. *Cell* **75** 1297-1303.

12   Pollak MR, Brown EM, Estep HL, McLaine PN, Kifor O, Park J, Hebert SC, Seidman CE & Seidman JG (1994) Autosomal dominant hypocalcaemia caused by a Ca^{2+}-sensing receptor gene mutation. *Nature Genetics* **8** 303-307.

196

The Comparative Endocrinology of Calcium Regulation
Eds C Dacke, J Danks, I Caple & G Flik, pp 197-199
Journal of Endocrinology Ltd, Bristol (1996)

Sun exposure and vitamin D synthesis and metabolism in two species of underground dwelling mole-rats

T Pitcher and R Buffenstein

Department of Physiology, University of the Witwatersrand Medical
School, 7 York Road, Parktown 2193, Johannesburg, South Africa
(T Pitcher is now at Department of Biological Sciences, 1.124 Stopford
Building, University of Manchester, Oxford Road, Manchester M13 9PT,
UK)

African mole-rats (*Rodentia*; *Bathyergidae*) naturally have an impoverished vitamin D status: they are strictly subterranean and fully deprived of sunlight; furthermore, they consume a vitamin D-deficient herbivorous diet. Plasma concentrations of vitamin D metabolites concur with vitamin D-deficiency (1); calcifediol (25-hydroxyvitamin D, 25(OH)D) is below the sensitivity of our assay (<5 nmol/l) while calcitriol (1,25-dihydroxyvitamin D, $1,25(OH)_2D$), although detectable (41±10 pmol/l), is lower than that reported for other rodents (3). Nevertheless, the vitamin D system in mole-rats is functional, as mole-rats convert vitamin D, if available, to more polar metabolites (2). We, therefore, questioned whether these animals can utilise conventional sunlight-mediated pathways to synthesise vitamin D and assessed changes in vitamin D status by monitoring changes in plasma vitamin D metabolite levels and hydroxylase activity following sunlight (UV) exposure or oral vitamin D_3 (D_3) supplementation.

Materials and methods

We compared two mole-rat species - *Heterocephalus glaber* (n=14), which has no fur, and the Damara mole-rat, *Cryptomys damarensis* (n=17), whose feet and nose are bare. Animals were housed under dim incandescent lighting (40 W). Animals were either exposed to sunlight (D_{UV}) over a 10 day period (*C. damarensis*, n=6; *H. glaber*, n=5) or given an oral D_3 supplement; two doses of D_3 were given: 1 IU/g dry matter food eaten (D_S) over a 10 day period (*C.damarensis*, n=5) or a single bolus of 42 000 IU (D_{SS}) 10 days before the animals were killed (*H. glaber*, n=4). After this time interval, animals were sacrificed, renal hydroxylase activities were determined *in vitro* using renal slices (7) and plasma vitamin D metabolite concentrations were measured by competitive binding assays (4, 6).

This study was approved by The Animal Ethics Committee of the University of the Witwatersrand, South Africa – Ethics number: 93/68/2b.

Results and discussion

In unsupplemented animals and those which were not exposed to sunlight (control, D_N), the elevated renal 25-hydroxyvitamin D-1α-hydroxylase (1-OHase) activity, coupled with the low plasma vitamin D metabolite concentrations (Table 1), confirm that mole-rats are naturally vitamin D-deficient. Exposure to sunlight leads to changes in hydroxylase activity similar to those induced by oral vitamin D supplementation (Fig. 1), indicating that mole-rats, like superterranean mammals (5), can endogenously synthesise vitamin D using UV-dependent photolytic pathways. Following 10 h of

Table 1 Vitamin D metabolite concentrations in mole-rats following sunlight exposure (D_{UV}), physiological oral vitamin D_3 supplementation (D_S), or a single supraphysiological vitamin D_3 dose (D_{SS})

	Cryptomys damarensis			*Heterocephalus glaber*		
	D_N	D_{UV}	D_S	D_N	D_{UV}	D_{SS}
25(OH)D (ng/ml)	< 5	< 5	30.78±2.22*	< 5	< 5	>100
1,25(OH)$_2$D (pg/ml)	12.09±1.30	23.35±2.86*	31.73±5.42*	17.2±5.28	27.1±2.65*	>210

Values are means±SEM. *$P<0.05$ compared with the control group (D_N).

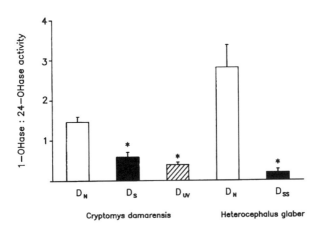

Fig. 1 Ratio of renal hydroxylase activity in mole-rats following sunlight exposure (D_{UV}), physiological oral vitamin D_3 supplementation (D_S) or a single, supraphysiological vitamin D_3 dose (D_{SS}). Values are means±SEM. *$P<0.05$ compared with the control group (D_N).

sunlight exposure 1-OHase activity is down-regulated and that of 25-hydroxyvitamin D-24R-hydroxylase (24-OHase) elevated. Graded responses to oral supplementation and sunlight exposure imply that the system is well regulated.

Despite down-regulation of 1-OHase activity following UV exposure, 25(OH)D plasma concentrations are still undetectable and even $1,25(OH)_2D$ (Table 1) is at the lower end of the normal range (3). No species differences were observed despite the greater skin surface area exposed to UV in *H. glaber*. Collectively, these data suggest that the vitamin D endocrine system in mole-rats is specifically down regulated and indeed, in these underground animals, may be vestigial.

Conclusion

Both *C. damarensis* and *H. glaber* appear to synthesise vitamin D and convert this prohormone to its active metabolite when exposed to sunlight. Living in a dark habitat, this pathway usually is not employed and even following the unusual event of prolonged sunlight exposure, endogenous vitamin D synthesis and metabolism is down-regulated.

Acknowledgements

We gratefully acknowledge assistance from Dr IN Sergeev, the MRC Mineral Metabolism Research Unit and in particular from Professor J Pettifor and Ms D Zachen, and funding from the MRC (South Africa).

References

1   Buffenstein R, Skinner DC, Yahav S, Moodley GP, Cavaleros M, Zachen D, Ross FP & Pettifor JM (1991) Effect of oral cholecalciferol supplementation at physiological and supraphysiological doses in naturally vitamin D_3-deficient subterranean Damara mole-rats (*Cryptomys damarensis*). *Journal of Endocrinology* **131** 197-202.

2   Buffenstein R, Jarvis JUM, Opperman LA, Cavaleros M, Ross FP & Pettifor JM (1994) Subterranean mole-rats naturally have an impoverished calciol status, yet synthesise calciol metabolites and calbindins. *European Journal of Endocrinology* **130** 402-409.

3   Chesney RW, Rosen JF, Hamstra AJ, Smith C, Mahaffey K & DeLuca HF (1991) Absence of seasonal variation in serum concentrations of 1,25-dihydroxyvitamin D despite a rise in 25-hydroxyvitamin D in summer. *Journal of Clinical Endocrinology and Metabolism* **53** 139-142.

4   Haddad JG & Chyu KJ (1971) Competitive protein-binding radioassay for 25-hydroxycholecalciferol. *Journal of Clinical Endocrinology and Metabolism* **33** 992-995.

5   Holick MF (1991) Photosynthesis, metabolism and biological actions of vitamin D. In *Nestle Nutrition Workshop Series*, vol 21, pp 1-22. Ed FH Glorieux. New York: Raven Press.

6   Reinhardt TA, Horst RL, Orf JW & Hollis BW (1984) A microassay for 1,25-dihydroxyvitamin D not requiring high performance liquid chromatography: application to clinical studies. *Journal of Clinical Endocrinology and Metabolism* **58** 91-98.

7   Sergeev IN, Arkhapehev P & Spirichev VB (1990) Ascorbic acid effects on vitamin D_3 hormone metabolism and binding in guinea pigs. *Journal of Nutrition* **120** 1185-1190.

The Comparative Endocrinology of Calcium Regulation
Eds C Dacke, J Danks, I Caple & G Flik, pp 201-203
Journal of Endocrinology Ltd, Bristol (1996)

Nucleotide sequence of canine preproparathyroid hormone

T J Rosol, L K McCauley[1], C L Steinmeyer and C C Capen

Department of Veterinary Biosciences, The Ohio State University, 1925
Coffey Road, Columbus, OH 43210, USA and [1]Department of Periodontics,
Prevention and Geriatrics, University of Michigan, 1011 N University
Avenue, Ann Arbor, MI 48109, USA

Introduction

The dog is an important animal in the study of normal and abnormal calcium balance
and calcium-regulating hormones *in vivo* (9). Serum concentrations of parathyroid
hormone (PTH) in dogs have been successfully measured using N-terminal and
two-site immunoassays to human PTH (hPTH) (6); however, the amino acid sequence
of canine PTH (cPTH) is unknown. The aims of this investigation were to: (i) develop
a lambda phage cDNA library to normal canine parathyroid glands; (ii) isolate and
sequence a cDNA clone for canine PTH; (iii) predict the primary protein structure of
cPTH; and (iv) compare the amino acid sequence of cPTH to human, bovine, porcine,
and rat PTH.

Materials and Methods

Complementary cDNA library: screening and sequencing

Parathyroid glands were removed from normal dogs, frozen in liquid nitrogen, and
stored at -80 °C. Polyadenylated (A$^+$) RNA was isolated from the tissue and a cDNA
library was prepared in lambda UNI-ZAP^{tm}XR (Stratagene, La Jolla, CA, USA) using
5 µg of poly(A$^+$) RNA. The cDNA library was screened with a hPTH cDNA (m124)
provided by Dr H Kronenberg (Massachusetts General Hospital, Boston, MA, USA)
(1) that labeled with [^{32}P]dATP by the random primer method. Approximately
600 000 plaques were screened and one lambda phage with a 692 bp insert (CPTH4)
was purified. A phagemid (pBluescript) containing CPTH4 was excised from the
lambda UNI-ZAP^{tm}XR. The 692 bp cDNA for canine PTH was sequenced in both
directions using an ABI 373A DNA sequencer. The ends of the CPTH4 cDNA were
sequenced using oligonucleotide primers to the T3 and T7 promoters of pBluescript
and the sequencing was completed using primers specific for CPTH4.

Comparisons to known sequences of PTH

The nucleotide and predicted amino acid sequences of cPTH cDNA (CPTH4) were
compared with human (V00597), bovine (J00023, K01938), porcine (X05722), and rat
(K01267, K01268) sequences obtained from GenBank or EMBL using Align Plus,
version 2.0 (2-4, 8, 10, 11).

Results and discussion

A 692 bp cDNA (CPTH4) which encodes for cPTH was isolated (Fig. 1) (7). The open reading frame of 84 amino acids (aa) was verified by the presence of a 5' in-frame termination codon (TGA) at bp 434-436. The partial 5'-UTR (bp 1-88) is 70, 68, 61 and 50% homologous to the corresponding region of bovine, human, porcine and rat PTH. The partial 3'-UTR (bp 437-692) is 80, 71, 71, and 59% homologous to the 3'-UTR of exon 3 of human, bovine, porcine, and rat PTH. There is 92, 91, 90 and 80% homology in the coding region of CPTH4 (bp 89-433) to bovine, porcine, human and rat PTH. In analogy to hPTH, there is a predicted cleavage site of the preproleader sequence from the mature peptide at amino acid 31. The predicted mature protein (aa 1-84) is 91, 87, 86 and 73% homologous to bovine, porcine, human and rat PTH (Fig. 2). There are only two substitutions in the first 40 aa of cPTH compared with hPTH at positions 7 and 16. These substitutions are not unique to cPTH since they also occur in bovine (position 7 and 16) and porcine PTH (position 16) (5). Knowledge of the sequence of cPTH will permit synthesis of peptides and development of immunoassays specific for the dog.

```
CGGCACGAGC ACAAGTTTAC TCAACTTCGA AAAAGCATCA GCTGCCGATA CACCTGAAAG   60
ATCTTGTCAC AAGACATTGT GTGTGAAGAT GATGTCTGCA AAAGACATGG TTAAAGTAAT  120
GATTGTCATG TTTGCAATTT GTTTTCTTGC AAAGTCAGAT GGGAAACCTG TTAAGAAGAG  180
ATCTGTGAGT GAAATACAGT TTATGCATAA CCTGGGCAAA CATCTGAGCT CCATGGAGAG  240
GGTGGAATGG CTACGGAAGA AGCTCCAGGA TGTACACAAC TTTGTTGCCC TTGGAGCTCC  300
AATAGCTCAC AGAGATGGTA GTTCCCAGAG GCCCCTAAAA AAGGAAGACA ATGTCCTAGT  360
TGAGAGCTAT CAAAAAAGTC TTGGAGAAGC CGACAAAGCT GATGTGGATG TATTAACTAA  420
AGCTAAATCC CAGTGACGAT ACATCAGGGC ACTGCTGTAG ACAGCATAGG GCAACAACAT  480
TACAAGCTGC TAACATTTTC AAGCTCTTAA GATTAATAAA TGCCCAAATT TACATGTAAT  540
CCATTGTTAG CCATGATAGC TGAAATTTTA ATTGATTGTT TTGATTCTAG TTTAATTCAT  600
TTAAGAGCTC TTTTAATTGT TCTATTTCTA TTGTTTATTC TTTTTAAAGT ATGTTTTTGC  660
ATAATTTATA AAAGAATAAA ATTGCACTTT TT                                692
```

Fig. 1 Canine PTH cDNA (CPTH4) (GenBank accession No. U15662).

```
cPTH  ( 1)   SVSEIQFMHNLGKHLSSMERVEWLRKKLQDVHNFVALGAPIAHRDGSSQR
bPTH  ( 1)   A.................................................S..Y.....
pPTH  ( 1)   .....L..........L........................S.V....G...
hPTH  ( 1)   .....L........N......................L.P..AG...
rPTH  ( 1)   A.....L........A.V..MQ.............S..VQM.A.E..Y..

cPTH  (51)   PLKKEDNVLVESYQKSLGEADKADVDVLTKAKSQ
bPTH  (51)   .R..........H.................I...P.
pPTH  (51)   .R..........H..........A....I...P.
hPTH  (51)   .R..........HE...........N.......
rPTH  (51)   .T...E....DGNS.....G........V.....
```

Fig. 2 Canine (c), bovine (b), porcine (p), human (h) and rat (r) PTH (1-84).

References

1   Born W, Freeman M, Hendy GN, Rapoport A, Rich A, Potts Jr JT & Kronenberg HM (1987) Human preproparathyroid hormone synthesized in *Escherichia coli* is transported to the surface of the bacterial inner membrane but not processed to the mature hormone. *Molecular Endocrinology* **1** 5-14.

2   Heinrich G, Kronenberg HM, Potts Jr JT & Habener JF (1984) Gene encoding parathyroid hormone: nucleotide sequence of the rat gene and deduced amino acid sequence of rat preproparathyroid hormone. *Journal of Biological Chemistry* **259** 3320-3329.

3   Hendy GN, Kronenberg HM, Potts Jr JT & Rich A (1981) Nucleotide sequence of cloned cDNAs encoding human preproparathyroid hormone. *Proceedings of the National Academy of Sciences of the USA* **78** 7365-7369.

4   Kronenberg HM, McDevitt BE, Majzoub JA, Nathans J, Sharp PA, Potts Jr JT & Rich A (1979) Cloning and nucleotide sequence of DNA coding for bovine preproparathyroid hormone. *Proceedings of the National Academy of Sciences of the USA* **76** 4981-4985.

5   Kronenberg HM, Bringhurst FR, Segre GV & Potts Jr JT (1994) Parathyroid hormone biosynthesis and metabolism. *The Parathyroids,* pp 125-138. Eds JP Bilezikian, R Marcus & MA Levine. New York: Raven Press Ltd.

6   Nagode LA & Chew DJ (1991) The use of calcitriol in treatment of renal disease of the dog and cat. *Proceedings of the 1st Purina International Symposium on Nutrition* **1** 39-49.

7   Rosol TJ, Steinmeyer CL, McCauley LK, Gröne A, DeWille JW & Capen CC (1995) Sequences of the cDNAs encoding canine parathyroid hormone-related protein and parathyroid hormone. *Gene* **160** 241-243.

8   Schmelzer HJ, Gross G, Widera G & Mayer H (1987) Nucleotide sequence of a full-length cDNA clone encoding preproparathyroid hormone from pig and rat. *Nucleic Acids Research* **15** 6740.

9   Slatopolsky E, Lopez-Hilker S, Delmez J, Dusso A, Brown A & Martin KJ (1990) The parathyroid-calcitriol axis in health and chronic renal failure. *Kidney International* **38** S41-S47.

10   Vasicek TJ, McDevitt BE, Freeman MW, Fennick BJ, Hendy GN, Potts Jr JT, Rich A & Kronenberg HM (1983) Nucleotide sequence of the human parathyroid hormone gene. *Proceedings of the National Academy of Sciences of the USA* **80** 2127-2131.

11   Weaver CA, Gordon DF, Kissil MS, Mead DA & Kemper B (1984) Isolation and complete nucleotide sequence of the gene for bovine parathyroid hormone. *Gene* **28** 319-329.

Part Four

Overview

The Comparative Endocrinology of Calcium Regulation
Eds C Dacke, J Danks, I Caple & G Flik, pp 207-209
Journal of Endocrinology Ltd, Bristol (1996)

Symposium summary and overview

D R Fraser

Department of Animal Science, The University of Sydney, N.S.W. 2006,
Australia

In reviewing the proceedings of this symposium I am reminded of an earlier meeting, in 1984, on the comparative endocrinology of calcium and phosphorus. This was part of the XVIIIth European Symposium on Calcified Tissues at Angers in France attended by about 400 participants. Nevertheless, during the comparative endocrinology session only about 20 people remained in the auditorium. Why is it that a small band of enthusiasts is enthralled by this field of comparative biology whereas for the majority of those working in calcified tissue research it apparently arouses little interest?

One possibility is that we simply find this topic fascinating and exciting: it provides an intellectual challenge which we enjoy. A more general reason, which we all believe, is that an understanding of comparative calcium homeostasis will give us a better insight into the mechanisms by which calcium and phosphorus are manipulated in human biology. However, my personal explanation for the fascination of this field is that it provides a splendid application of the theory of evolution - a theory which cannot be proven but which is remarkably useful to applied biologists.

Evolutionary biology solved the problems of handling calcium a long time ago. Consider the early amphibian *Eryops,* a large animal compared with the modern frog, being about 1.5 m in length. *Eryops* had massive membranous bones of the skull and massive limb bones formed by endochondral ossification. The problem of constructing a weight-bearing skeleton for very large land animals had clearly been overcome during the early stages of terrestrial vertebrate evolution. However, the last *Eryops* disappeared about 300 million years ago.

As well as in the mighty dinosaurs, gigantic bone structures were found in the evolving mammals. The largest terrestrial mammal was *Baluchatherian,* a giant Asian rhinoceros standing about 6 metres high. *Baluchatherian* can no longer be found, the genus having died out about 4-5 million years ago.

All the mechanisms for obtaining and utilising calcium and phosphorus in the construction of mineralised tissues as well as for diverse cell functions had become established at a very early stage in animal evolution. The difficulty for us, in learning about those mechanisms, is that all the species in which evolutionary developments in mineral metabolism occurred, are now long extinct. Thus, comparative endocrinology of calcium and phosphorus metabolism is really a forensic science and as in all forensic investigations we have to draw conclusions and make judgements on imperfect evidence. It is therefore necessary to make imaginative leaps to try to fill the gaps in our imperfect understanding.

Hence, in trying to summarise this symposium, I find myself a bit like a coroner at an inquest, attempting to sum up a body of incomplete evidence on extinct life-forms so that you, the jury, can reach a verdict.

The fundamentals of calcium metabolism are quite simple. They are the transport processes which vary the concentration of calcium on opposing sides of various cell membranes. The components include: (a) ion-specific channels; (b) ion-specific pumps; (c) regulatory factors that stimulate cells to use the channels or pumps; (d) receptors for those regulatory factors.

The regulatory factors considered in this symposium include: (a) parathyroid hormone (PTH); (b) parathyroid hormone-related protein (PTHrp); (c) calcitonin; (d) stanniocalcin; (e) vitamin D metabolites

To improve our understanding of how these all work we need to draw conclusions about: (a) the mechanisms of calcium transport across membranes; (b) the role of various extracellular regulatory factors on the membrane transport of calcium

At this symposium we have heard evidence about the action of calcium pumps localised in the basolateral membranes of polar cells in the intestinal mucosa, the nephron and in gill filaments. Allied to these, we have heard about the synthesis and properties of the mysterious intracellular calcium-binding protein (CaBP). This protein, which was discovered 30 years ago by Wasserman and his colleagues, is now characterised by an extensive catalogue of information in many species, from the skilled research of this Cornell Group. However, are we yet certain about the biological role of this protein in the cellular transport of calcium? Is there a functional relationship between CaBP closely associated with calcium ion pumps in plasma membranes?

The evidence is now quite convincing that PTH and PTHrP are ancient peptide hormones which evolved long before the appearance of land vertebrates with their discrete PTH-secreting glands. The original function of these hormones may not have been in calcium homeostasis but in neurotransmission. Is a neurotransmitter role still a major function of these peptides in terrestrial vertebrates? It appears that one of the responses to PTH by neurones from various vertebrate and invertebrate species is a rapid change in cytoplasmic calcium concentration. Could this action in neurones be the origin of the PTH role in modern calcium homeostasis? Is the action of PTHrp in fetal calcium homeostasis a demonstration of the transition between the ancient and modern roles of these peptide hormones?

In contrast to PTH, a hormone discovered in tetrapods, the peptide hormone, stanniocalcin, was extracted from glandular tissue found only in teleost fish. This hormone has a well-defined role in fish calcium homeostasis. Its action on cells in the gill filaments blocks the uptake of calcium across the apical cell membranes. Has this hormone been lost in the land vertebrates or does it still exist with a function derived from its inhibitory role on calcium transport in the gills of teleost fish?

Exploration of the comparative physiology of calcium fluxes *in vivo* continues to demonstrate tantalising phenomena with imaginative explanations. In an ingenious analysis of the remarkable capacity of bone to clear circulating calcium, Bronner has concluded that exposure of calcium affinity sites on the crystalline surfaces of bone may be regulated by the physical contraction or expansion of overlaying osteoblasts under the influence of PTH or calcitonin. Hurwitz has shown in fast-growing chickens that the integrated controls of calcium homeostasis cause oscillations in plasma calcium concentration. One of the homeostatic mechanisms could be mediated by the recently discovered calcium-sensing receptor on osteoclasts. Osteoclast behaviour is modified by changes in extracellular Ca^{2+} concentration detected by this calcium receptor.

An emerging puzzle: how does vitamin D have a central role in whole body calcium homeostasis at the same time as a multitude of apparently unrelated actions in most cell types? Perhaps studies of vitamin D physiology in fish may reveal a simpler role than the functional complexity found in mammals and birds. One finding is that the 1α-hydroxylase is located mainly in the liver of fish. Does the relocation of this enzyme to the kidneys of tetrapods indicate a change in the significance for calcium homeostasis of $1,25(OH)_2D$ and of the kidney with the adaptation of vertebrates to life out of water?

It has gradually been accepted, over the last 25 years, that vitamin D is the precursor of a classical steroid hormone, acting in cell nuclei on the expression of genomic information. Now, evidence for non-genomic functions in cell membranes is appearing. Have we yet got a conceptual framework for vitamin D which really explains its function? I am reminded of the intuitive explorations of Kodicek in attempting to find a mechanism of action for vitamin D. He demonstrated (1) that vitamin D, at low concentration, had a specific effect on the growth of Gram-positive bacteria. These early studies in comparative biology may have revealed a property of the vitamin D molecule in biological membranes which could be linked to the recently described non-genomic effects of $1,25(OH)_2D$.

The comparative research described in this symposium has caused us to re-evaluate our views on the regulation of calcium metabolism. The novel questions that arise from these investigations would not have been provoked if we had not tried to understand the development of these control mechanisms during evolution of animal biology. Hopefully, the publication of these findings will persuade others interested in the biology of bone and mineral that the comparative approach enriches our understanding of human biology.

Reference

1   Kodicek E (1956) The effect of unsaturated fatty acids, of vitamin D and other sterols on Gram-positive bacteria. *Biochemical Problems of Lipids*, pp 401-406. Eds G Popjak & E Le Breton. London: Butterworths.

Author index

Subject index

Subject index

Subject index

marine animals
  salt and water balance
    extracellular solute composition 18-24
    osmolytes 32-3
    perturbing, compatible and couteracting
      solutes 34-6
    positive buoyancy solutes 32-3
    reinvasion of seawater 25-8
    seawater-brackish-freshwater transitions
      24
    terrestrial transitions 25-8
medullary bone turnover 143-8
metabotropic glutamate receptors 193-4, 195
mole-rats
  calcium homeostasis 177-80
  vitamin D system 177-80, 197-9
molluscs
  salt and water balance
    extracellular solute composition 19, 21,
      22, 23
    perturbing, compatible and counteracting
      solutes 35
    positive buoyancy solutes 32
    seawater-brackish-freshwater transitions
      24
    terrestrial transitions 28-9

neural Ca^{2+} regulation 7-10
N-terminal immunoassay 185-6, 187, 192, 194
nucleotide sequencing
  canine preproparathyroid hormone 201-2
  canine PTHrP 186-7
nude mice 184-5

onycophorans
  salt and water balance
    extracellular solute composition 19, 22
    terrestrial transitions 29
osmo-regulation
  salt and water balance
    aquatic-terrestrial transitions 28-32
    extracellular/intracellular fluid
      composition 17-24
    osmolytes 32-3
    perturbing, compatible and counteracting
      solutes 34-6
    positive buoyancy solutes 32-3
    reinvasion of seawater 25-8
    seawater-brackish-freshwater transitions
      24

osteoblasts
  avian bone turnover 113-14
    role of estrogen 117
    role of PTH 114-17
  effects of the vitamin D_3 system 78
  hormone interaction with cell receptors
    132-4
  plasma osteocalcin 158, 160, 161
osteoclasts
  avian bone turnover
    Ca^{2+}-sensing receptors 123-8
    changes in medullary bone during
      egg-laying cycle 149-50
    formation of resorption pits in medullary
      bone 143-8
    role of calcitonin 117-18
    role of PTH 115-17
  hormone interaction with cell receptors 132
osteoporosis 154

parathyroid hormone (PTH)
  avian bone turnover 114-17
  calcium metabolism 132-4, 157
  evolution of neural PTH
    blood-brain barrier 4
    ectodermal origin 3
    ectopic transcription of PTH genes 4-5
    evolution of the PTH gene 5-6
    neural roles 7
    occurrence in tetrapods 3-4
    PTH-like peptides 5
    whole-body Ca^{2+} homeostasis 6-7
  evolutionary biology 208
  fetal calcium homeostasis 159-60, 162, 163
  humoral hypercalcemia of malignancy in
    dogs 185-6
  inhibition of ruminal motility 171-5
  neural Ca^{2+} regulation 7-10
  response of the calcium-regulating system in
    chickens to growth 137, 138 140
  Xenopus laevis 97-100
parathyroid hormone-related protein (PTHrP)
  157
  evolutionary biology 208
  fetal calcium homeostasis 159-60, 162, 163
    renal calcium excretion in the ovine fetus
      165-8
  fish
    corpuscles of Stannius 107-8